流行中式點心

茶粿、酥餅、糕點、包子饅頭 一次學會

獨角仙 著

自 序

　　當編輯邀請我出版一本中式點心的食譜書，我一口答應！雖然這和我做麵包是風馬牛不相及的事，但我多年出外學習和交流下，發覺做料理是不可以執著於某國菜式和做法，好像以前學西點的我會轉向做麵包，更以過去所學的甜點概念融入我的麵包中，而中式點心亦啟發了我不少做麵包的靈感。如果說音樂是世界語言，飲食更是世界溝通的橋梁，相信沒有人是不喜歡吃的。

　　中式點心有著源遠流長的飲食文化背景，所屬派系亦五花八門，不同地域有不同的糕點小吃，各具特色。香港比較幸運，很多民族鄉里都滙聚在此，以客家、潮汕、京滬、淮揚、廣州影響尤甚。

　　雖然香港沒有生產米糧，但地利因素，進口的原材料極為廣泛，像植物性原材料如五穀、蔬菜、乾果，以及動物性原材料，如禽畜類、水產類，這些材料都是我們日常生活經常食用的。點心的製作，只是將這些食物給予重新組合，成為主食或主副食兼具的包點，也有成為應節食品。

　　餅點製作所使用的原材料廣泛，是由於各地區風味不同，物產不同所至，所以我們對原材料選用要有基本的知識，了解各種原料的種類、性能和用途、原料的加工和處理方法、使用方法、配合方法，才能做出色香味形質俱全的食品。在這書中我將其主要歸納在四大種類：茶粿、酥點、糕點、包子饅頭，包括了地道的廣式餅點和京滬餅點。

Preface

When the editor asked me to write a book on Chinese Dim Sum, I said yes without hesitation. I'm actually a baker and Chinese Dim Sum doesn't seem to be my business at all. But after years of learning and working, I found a real culinary artist must not be restricted by any geographical or categorical boundary. I started out as a pastry chef, but I switched to bakery later on. The knowledge I acquired in a patisserie didn't go wasted. In fact, I always incorporate some pastry skills in my bread. Another major influence on my bakery is Chinese Dim Sum. People say music is the universal language. I'd say food is the universal bridge of communication. Good food speaks to anyone and everyone, across all races.

Chinese Dim Sum has a profound culinary history and there are countless variations from different parts of China. Every county has its own signature sweets and snacks. As a hotchpotch of cultures, Hong Kong is valued for its eclectic mix of quality Chinese food, including Hakkanese, Chaoshan, Beijing, Shanghainese, Huaiyang and Cantonese cuisines.

Hong Kong doesn't grow its own grain crops. But its central location favoured the importation of raw materials from all over the world. We consume botanical produce such as grains, vegetables, fruits, and animal produce such as poultry, meat and seafood on daily basis. The same raw materials are also put together creatively in Dim Sum making, for buns and snacks that serve either as staples or as a combination of staples and non-staples. Certain recipes also make great festive munchies.

Countless ingredients are used to make cakes and snacks. It comes down to the regional palate and availability of certain ingredients. Before we can make snacks and cakes with authentic taste, smell and look, we must master the basic knowledge on ingredients; understand their categories, properties and uses; how they are processed and used, and their classic combinations. I categorized the recipes in this book into 4 sections – homestyle dumplings, Chinese puff pastries, cakes and buns. Most of them are of either Cantonese or Beijing/ Shanghainese origin.

Kin Chan

推薦序

一代接一代、
延續自然美味的中式點心

　　談到中式點心，就想起小時候老母親的「炸鹹芋丸」，每逢過年及中元，老母親總會做出這一道點心來祭拜，在當時戰後物質不豐富的年代，父母總會想盡各種方式來餵飽孩子，將芋頭蒸熟、加入絞肉，並用紅蔥頭切碎油炸，加入裡面味道特別香。中式點心就是這種揉合著記憶、嗅覺、視覺與自然食物美味的綜合體，這麼地令我魂牽夢縈。

　　記得多年前曾看過一部電影《吐司》，這是改編自英國名廚奈傑史萊特的（Nigel Slater）的同名自傳（Toast: The Story of a Boy's Hunger）同名回憶錄。電影主角利用美食抓住父親的胃不只是個手段，後來更成為與後母一較高下的利器，二人互相較勁的一道點心「檸檬蛋白霜派」令我印象深刻。對我來說，美味的點心是視覺、聽覺的雙重享受；就像「炸棗」時；油在鍋內輕快彈跳的聲音、「烤杏仁餅」時；烤箱滋滋作響的呼喚、攪拌麵糊時；淫搭黏稠的聲音，種種的視覺和聽覺，總喚起我對中式點心最心動的記憶。

　　在色彩繽紛、外型多變的點心世界中；中式點心顯得較為刻板，圓就是圓，扁就是扁，鮮少有變化。當我得知《無包不歡》榮獲 2013 世界美食家圖書大獎的作者獨角仙（仙姐），即將在台灣

出版這本《流行中式點心》時，感到非常地雀躍。她將「中式點心」融入了變化，啟用了創新，在每道甜點中運用了麵包的做法，巧妙的做出各種不同的風味點心。如同作者所言：「假如音樂是世界的語言，飲食更是世界溝通的橋梁」。 相信沒有人是不喜歡吃的。作者居住在香港，利用自然食材來製作各式點心：茶粿、酥點、糕點……，內容豐富，值得大家一窺究竟。希望讀了這本書的讀者，也能做出屬於自己的中式點心，讓中式點心得以一代接一代的繼續承傳下去。

中式點心糕餅職人、國寶級糕餅師傅
《懷舊糕餅 90 道》、《懷舊糕餅 2》 作者──呂鴻禹師傅

目錄

Contents

認識中式餅點

在這本書中，獨角仙為你介紹數十款中式餅點的製作方法和製作的技巧，有鄉土味濃的茶粿、酥鬆的酥點，新穎與傳統並重的糕點和包子、饅頭。在走進獨角仙的餅點國度前，讓我們先了解中式餅點常用的原材料、麵團分類、調製方法和餡料製作吧。

中式餅點常用的原材料

麵粉

麵粉根據蛋白質含量可分為超高筋、高筋、中筋、低筋四大類：

超高筋麵粉

含蛋白質量 15% 以上，麵粉吸水量高，口感有彈性，適合麵條的製作。

高筋麵粉

蛋白質含量約在 11.5-14.5%，口感有彈性，適合製作生煎包、葱油餅、水餃、雲吞皮等點心。

中筋麵粉

蛋白質含量為 9-10%，吸水量適中，可塑性高，韌性好，常用來製作包子、饅頭等點心。

低筋麵粉

蛋白質含量約 7.5-8.5%，口感幼滑，成品潔白細緻，可製作廣式包點、刀切饅頭、蛋糕等等點心。

糯米粉

糯米粉用圓糯米加工製成，黏度高，以柔軟、韌滑、香糯而著稱，它可以製作湯圓、年糕等。

在來米粉

在來米粉是用白米磨成的粉，又叫粘米粉，是多種食品的原料，是各種白米中糯性最低的品種，多用來製作蒸製的糕點，質地較鬆散，例如蘿蔔糕、芋頭糕等。

澱粉

澱粉是製作餅點的另一重要原材料，有多種來源，隨着植物不同，可區分為豆穀類、薯類、蔬菜類三大類。常用的有**小麥澱粉**（無筋麵粉）、**玉米澱粉**（粟粉）、**甘薯澱粉**（地瓜粉）、**木薯澱粉**（木薯粉）、**馬鈴薯澱粉**（太白粉）、**荸薺粉**等。

小麥澱粉也叫汀粉、澄麵，是將高筋粉麵團洗水搓揉，去掉麵筋，經沉澱後得出的澱粉。這澱粉口感爽口，沒有筋性，因為無筋麵粉黏度

和透明度均較高,蒸熟後看起來晶瑩剔透,主要用作製作中式糕餅或點心的粉皮,如廣式點心中的蝦餃、粉果的粉皮。

玉米澱粉、太白粉、木薯粉、荸薺粉多用於製作餡料,凝固或勾芡,亦可以配合其他粉料,作為改善麵團性質的材料。

油脂

用來調製麵團,改善麵團性質,使成品柔軟可口,可調製餡料,又是煎炸最主要的傳熱原料,常用的有固體豬油、奶油、植物油。

糖

糖在中式餅點中主要功用是增加甜味,調節口味,改善色澤,保持成品柔軟度和延長保存期。普遍使用的有白砂糖、糖粉、糖漿、紅糖、黑糖、冰糖等。

蛋

蛋的用途很多,可作餡料或油炸麵糊的黏結和增加香氣,也可增加成品的色澤和滑韌性,增加營養價值。常用的有雞蛋、鹹蛋和皮蛋等。

酵母

酵母是一種活性膨鬆劑,發酵力強,使產品膨大鬆軟。

發粉

又稱泡打粉,是一種膨鬆劑,使產品膨大鬆軟。

水

主要調節點心、麵團或麵糊的軟硬度。

食鹽

主要功用是提供鹹味,調節產品的味道,降低麵糊焦化度。

中式餅點的麵團分類和調製方法

調製麵團是將雜糧粉類如麵粉、米粉或其他雜糧粉，加入適當的水、油脂、蛋等材料後加以搓揉，調製，使其互相黏連形成一整體，經和麵、揉麵兩個過程，按不同烹調方法，例如蒸、煮、烙、煎、炸、烤等，做出粥、麵、糕、餅、團、粉、條、塊、卷、包、餃、羹、凍等不同形態的餅點。能做好這些，對掌握調製麵團的技術極為重要。

麵團通常可分為五大類：

無筋麵團

是以熱水倒入粉團或烹煮粉漿來調製，特點是色澤潔白呈半透明，細膩柔軟，入口嫩滑。成品如雜糧糕、馬豆糕、缽仔糕等。

油酥麵團

是指以油脂和麵粉為主要原料來調製麵團，有時再添加蛋、砂糖、化學膨脹劑。

大致可分為層酥、單酥、炸酥三類。層酥成品能看出層次，有明酥和暗酥之分，特點是有層次，入口鬆酥，色澤美觀，口感酥香。成品如：叉燒酥、皮蛋酥、合桃酥、芋頭酥等。

膨鬆麵團

是在調製過程中，加入酵母或化學膨脹劑等，使麵團有反應，變得膨脹疏鬆，特點是口感鬆軟，形態飽滿，營養豐富，易被人體消化吸收。成品如：饅頭、壽桃包、銀絲卷等。

水調麵團

是指用水和麵粉調成的麵團，不經過發酵，用水和麵粉直接搓揉成的麵團。分冷水麵團、熱水麵團和溫水麵團，用於製作各式麵點，用途最廣，成品如：生煎包。

米粉麵團

是指用米粉類為主材料，加入水、調味和油拌成粉漿，然後採用蒸製後成形，或成形後蒸製的方法做成品，如：蘿蔔糕、茶粿。

餡料製作

餡料按口味分鹹、甜兩種，按原料分為葷、素兩種，按製法可分為生餡、熟餡兩大類。

乾餡料要先泡軟或蒸軟才可使用，若帶有不良氣味、腥味、苦味或澀味就要先去除，水分多的，例如菜類，要先醃漬去掉多餘的水分。

餡料黏性會影響成品的成形和品質，素餡生拌時要減去原料的水分來增加黏性；生肉餡要加水或加入肉皮凍使餡料有黏性；熟餡黏性差，可用打芡方法增加黏性。亦可以油脂、蛋或醬料增加黏性，拌合時要快而均勻，防止餡料出水。另外也要注意材料所切的大小，口味的濃淡，乾濕來配合麵團。

茶粿

茶粿是客家人最常做的民間小吃,它有很多種類,大致分鹹、甜兩種口味。鹹茶粿餡料有蘿蔔、眉豆(又稱白豆、米豆)、蝦米、肉鬆;甜茶粿有豆沙、花生、芝麻、番薯、南瓜、雞屎藤等,還有其他特別口味,如喜粄(客家食品,家有初生嬰兒時,做喜粄送給親友,藉此告知喜獲麟兒)、茶粄等等,軟糯Q彈。

過去很少人賣茶粿,因為製作工序繁複,已經很少年輕人肯學,只有少數老攤有擺賣,香港這幾年來,在新界和離島遊客人數增多,而茶粿則是當地代表食物,所以多了人做來賣,也因此創造了很多新口味。

獨角仙很欣賞中式茶粿這些民間小吃,喜歡吃固然是原因,但最高興的事是做這些小吃時,總會一家大小圍著做,多數是節日前用來拜神,想像婆孫三代一起做茶粿、過年節,那是多麼美麗的畫面呢!做茶粿時,會用蘋婆葉做墊葉,記得我外婆家有一棵很大的蘋婆樹,向上望時,一纍纍火紅的果莢掛在樹上,那是蘋婆果(又稱鳳眼果)紅紅果莢內的果子,蘋婆果肉吃起來綿密粉嫩,有點像吃栗子,將它用水蒸煮享用,或是用來燜雞都可口。

不過吃蘋婆果是很講究技巧的,要先把紅色果殼內的咖啡色果實取出,可用手或小刀將咖啡色果殼外的軟皮、硬果殼、包著肉的軟皮剝開,黃澄澄像蛋黃的果肉就在眼前,用水蒸煮加入少許鹽把果殼蒸至微爆開,就代表熟了。

可是現在蘋婆葉比較難找到,都是用蕉葉或粽葉代替了。

南瓜紅豆
茶粿

約可做
16個

自家製的紅豆餡甜味適中，南瓜香甜口感綿密，
這茶粿是會吃上癮的。

材料

紅豆餡料

熟紅豆粒（看紅豆餡做法 1～3）	300 克
水	適量
砂糖	100 克
固體豬油	30 克

皮料

熟南瓜泥（日本南瓜）	450 克
糯米粉	240 克
在來米粉	40 克
紅糖	60 克
豬油	15 克
水（調節用）	少許

墊葉

蘋婆葉、蕉葉或粽葉	適量

做法

紅豆餡

1 將生紅豆浸軟,放入鍋裡,加水,用大火煮約 30 分鐘後,倒掉水分,這個步驟可去掉紅豆的苦澀味。

2 將紅豆再放回鍋子裡,加水,用大火煮至紅豆開始變軟,倒掉水分。

3 再加水,轉小火繼續煮至紅豆全部破開,倒掉水分。

4 在鍋內放入已煮破開的紅豆,加入砂糖和固體豬油不斷翻炒至收乾水分,圖 ①～②。

5 如果喜歡較綿的紅豆泥,可邊煮邊壓成泥,冷卻備用。

皮料

1 蘋婆葉、蕉葉或粽葉用水燙一會,撈起並剪成所需的形狀。

2 將糯米粉、在來米粉、紅糖、豬油放在大碗內。

3 南瓜蒸熟壓成泥狀,趁熱倒入步驟 2 有糯米粉、在來米粉等料的大碗內,用木棒攪勻如圖 ③,並搓勻成粉團 ④(可按南瓜水分含量多寡才加水調節粉團)。

4 可以拿一個圓形的器具,先將粉團分成 50 克,利用圓形器具塑成圓形,再包入紅豆餡 30 克如圖 ⑤～⑥,放在已塗油的墊葉上,茶粿表面則塗上薄薄的油如圖 ⑦。大火蒸約 13 分鐘即可。

Tips

1　要注意南瓜的含水量，中國南瓜的含水量較高，水用量可按粉團乾濕度增減。

2　九至十一月為日本南瓜的盛產期，南瓜會最便宜、最漂亮，但要看清楚是否日本原產地，因為市面有很多別國產的日本種南瓜，味道和品質就遜色多了，獨角仙買過墨西哥產和紐西蘭產的品質都可以，其他就要小心上當。

3　南瓜買回來後多放幾天，讓它自然熟成會更香甜，顏色更亮麗。

眉豆茶粿

約可做
16個

客家鹹茶粿的餡料並不複雜，採用簡單的食材，
便能創造出美味的小吃。

材料

餡料

乾眉豆 ....300 克
水 ....適量
乾蝦米 .... 50 克
油 .... 2 湯匙
紅蔥頭 .... 少許

皮料

糯米粉 ....320 克
在來米粉 .... 55 克
海鹽 .... 2 克
固體豬油 .... 30 克
熱水 ....300 克

調味料

海鹽 .... 8 克
砂糖 ....10 克
五香粉 .... 5 克
胡椒粉 ....少許
麻油 ....少許

墊葉

蘋婆葉、蕉葉或粽葉 .... 適量

裝飾

水溶性食用色素 ....適量

做法

餡料

1 先將乾眉豆用水煮約 40 分鐘至軟。

2 乾蝦米洗淨，浸軟切碎。

3 起油鍋，爆香紅蔥頭、蝦米後，下眉豆和調味料拌勻 ① 。

皮料

1 蘋婆葉、蕉葉或粽葉用水燙一會，撈起，並剪成所需的形狀。

2 將糯米粉、在來米粉和海鹽混合好。

3 加入熱水、固體豬油搓成粉團。

4 粉團分成 50 克，包入餡料 ② ，封口，造型 ③ ，放在已塗油的葉上 ④ 。

5 茶粿表面抹油，隔水蒸 13 分鐘。

6 用食用色素點在茶粿上裝飾 ⑤ 。

Tips

眉豆不要煮得太爛，炒完後還看到原本顆粒，軟軟的粉粉的才好吃。

甜菜根番薯茶粿

約可做
10-12
個

記得小時候外婆的田種著豆角、冬瓜，隔壁「惡婆婆」的田種著番薯，我們坐在田畦上伸手去隔壁的田挖番薯，順著番薯苗挖下，大的摘下來，小的放回去，生怕被「惡婆婆」罵，吃著這茶粿，想起很多。

材料

番薯餡料

熟番薯	250 克
砂糖	20 克
無鹽奶油	30 克

皮料

熟番薯泥	160 克
熟甜菜根泥	160 克
糯米粉	220 克
紅糖或黑糖	120 克
水	50 克
豬油	10 克

墊葉

蘋婆葉、蕉葉或粽葉	適量

做法

1. 將全部番薯餡料混合 ① ～ ② 。
2. 蘋婆葉、蕉葉或粽葉用水燙一會，撈起，並剪成所需的形狀。
3. 生番薯、生甜菜根連皮一起蒸熟。
4. 番薯去皮，過篩壓成泥；甜菜根去皮，刨成泥。
5. 將紅糖或黑糖與水一起煮溶，趁熱倒入糯米粉、熟番薯泥和熟甜菜根泥 ③ ，用木棒攪勻 ④ 。
6. 搓勻成粉團，再加入豬油搓勻 ⑤ 。
7. 按模具的大小，取適量的皮料，包餡 ⑥ ～ ⑦ 或不包餡均可，在模內撒上糯米粉，拍出多餘的糯米粉，按壓成型 ⑧ ，放在已塗油的葉子上 ⑨ 。
8. 大火蒸約 12 分鐘。

混合所有番薯餡料。

將糯米粉、熟番薯泥、甜菜根泥倒入糖水中攪勻。

加入豬油搓成團，分成適當大小包入餡料。

包好餡料後,放入模具內按壓成型。

放上已塗油的墊葉。

Tips

1 甜菜根品質不同,做出來的茶粿顏色會有深淺之別。

2 蒸茶粿時,要將不同大小的茶粿分開蒸,時間亦不宜過久,否則造形茶粿的花紋
  會因過度膨脹而變形,就不美觀了。

3 亦可以不用模具,按喜愛製作傳統的圓形茶粿。

4 不要一次倒下全部糖水,要按番薯泥和甜菜根泥的含水量來調整糖水用量。

5 可用有薑味的紅糖,風味更佳。

蘿蔔茶粿

約可做
12個

「冬吃蘿蔔、夏吃薑，一年四季保安康」，白蘿蔔是秋冬時令蔬果，含有豐富的營養，可提高免疫力，促進新陳代謝的功能，並有助於防癌。香港西貢區人喜歡自家種蘿蔔，留待過年做糕和茶粿，蘿蔔茶粿是地道美食，趕緊試試！

材料

餡料

白蘿蔔	300 克
紅蔥頭	適量
臘腸	45 克
乾蝦米	15 克
甜菜脯	15 克
蔥	適量

皮料

熱水	310-315 克
糯米粉	330 克
在來米粉	90 克
豬油	20 克
海鹽	3 克

調味料

砂糖	6 克
雞粉	2 克
海鹽	適量
胡椒粉、麻油	少許
油	適量
太白粉	適量
豬油	15 克

墊葉

蘋婆葉、蕉葉或粽葉	適量

30

做法

餡料

1 白蘿蔔去皮刨細絲，乾蝦米泡水浸軟，紅蔥頭、臘腸、甜菜脯切丁。

2 將白蘿蔔絲汆燙至軟，瀝掉水分。

3 起油鍋，爆香紅蔥頭、臘腸、蝦米和菜脯 ①，下蘿蔔絲、除了太白粉、豬油以外的調味料，爆炒至蘿蔔入味。

4 用太白粉勾芡，再放入豬油炒勻，放大碗內加蔥粒拌勻，放涼備用 ②。

皮料

1 蘋婆葉、蕉葉或粽葉用水燙一會，撈起，並剪成所需的形狀。

2 糯米粉、在來米粉篩勻，加入熱水 ③ 加入海鹽與豬油，搓成不黏手麵團，要注意按麵團軟硬加減水量 ④。

3 麵團分成 60 克，捏成窩形 ⑤，包入 35 克餡料 ⑥，捏緊收口，收口向下，輕壓扁，放在已塗油的葉子上 ⑦。

4 以大火蒸約 15 分鐘即成。

Tips

1 皮料的糯米粉和在來米粉比例，可按個人喜好調整。
2 傳統的客家人用木球按入麵團成窩狀，書中所用的是雲石球（請見圖 ⑤），可視自己方便使用不沾黏材質的球狀體。

①

②

雞屎藤茶粿

約可做
27個

香港最具特色的茶粿，非雞屎藤茶粿莫屬，它屬茜草科，別名臭腥藤，味甘、微苦，性平，是香港常見的植物，亦是山草藥，功能祛風活血，散瘀消腫，清熱解毒，止咳止痛，外敷內服皆宜。客家人就地取材，以其葉子製汁做茶粿，通常在清明前後才會做，屬於季節性食品。雞屎藤的葉子帶著雞屎臭味，但烹煮後會散發出一種獨特的清香，吃罷清熱解毒，是保健食品。

材料

餡料
去殼花生	80 克
黑芝麻	10 克
白芝麻	15 克
紅糖	40 克

皮料
雞屎藤葉	200 克
水	500 克
片糖	80 克
糯米粉	500 克
在來米粉	60 克
豬油	30 克

墊葉
蘋婆葉、蕉葉或粽葉	適量

做法

餡料

1 黑、白芝麻洗淨。
2 花生仁、芝麻烘香後攪碎 ①，加入紅糖成餡料 ②。

皮料

1 蘋婆葉、蕉葉或粽葉用水燙一會，撈起，並剪成所需的形狀。
2 雞屎藤葉洗淨，加水用磨砵 ③~④ 或攪拌機打碎，連渣加入片糖煮滾至片糖融化，稍微放涼備用 ⑤。
3 粉類放入大碗內與雞屎藤糖水混合 ⑥，加入豬油搓成團，要注意雞屎藤糖水份量，要視麵團軟硬加減 ⑦。
4 麵團分成 50 克一份，包入餡料 ⑧~⑩，捏緊收口，收口向下。
5 輕輕壓扁，放在已塗油的葉上，茶粿表面塗上油 ⑪。以大火蒸約 15 分鐘即成。

Tips

雞屎藤葉可在郊外地方採摘，也可在新鮮山草藥店舖找到，但要小心挑選，雞屎藤葉分有毛面和光面兩種，客家人只會用光面的葉造茶粿。製作茶粿的墊葉多是就地取材，如蕉葉、粽葉或蘋婆葉（鳳眼果葉），今次我用上蘋婆葉。蘋婆葉的葉子大，葉面光滑，耐久蒸不爛，而且有一股清香，又不會喧賓奪主。

搓成團時，要視麵團軟硬調整加入的糖水份量。

利用工具做出窩型，包入餡料。

表面塗上油即可。

茶粿
不只是好吃的中式點心，
更是連結家族情感的動人滋味。

酥點

傳統的中式酥皮，分成淮陽酥皮和廣式酥皮兩大類別。

淮陽酥皮類，像是蟹殼黃、菊花酥、叉子燒餅、胡椒餅、蘿蔔酥等等，廣式酥皮類，則是像嫁女餅紅綾、白綾、黃綾（類似台灣傳統喜餅）、皮蛋酥、老婆餅等等。酥皮做法分大包酥和小包酥，大包酥做法類似做西式酥皮，大水皮包入大酥心，優點是速度快、效率高、可大量製作，缺點是不容易擀均勻、油酥層次少，酥鬆性差。自家製作常用的做法是小包酥，水皮、酥心分割成小麵團後分別包製、擀捲，優點是容易擀捲，層次清晰，酥鬆性好，缺點是速度慢、效率低，不適合大量製作。

廣式酥皮是油酥麵團之一，使用油脂、麵粉、雞蛋和砂糖等原材料，經複雜的手工製作而成，特點是酥層酥鬆、色澤美麗、味道豐富。淮陽酥皮均是油酥麵團，依酥皮的形成，可分為明酥型、暗酥型、半暗酥型與發麵油酥。

明酥型：表面可以看到清楚的紋路及層次；暗酥型：表面看不出層次，須切開後，從橫切面才看得到。半暗酥型：酥層一部分在裡面，一部分在外面。發麵油酥：利用發麵類麵團包裹油酥，經皺摺而產生層次。

中式酥皮和西式酥皮的起酥原理都是一樣的，都是於麵團中裹入油脂，經過反覆摺疊，形成數百層。麵皮和油脂的分層在烤焗的時候，麵粉顆粒被油脂顆粒包圍隔開，使得麵粉顆粒間的距離擴大，空隙中充滿了空氣，空氣受熱膨脹，使成品疏鬆，形成層次分明又香酥可口的酥皮。

廣式酥皮的製作關鍵在於麵團的調節和擀麵的技巧。首先水皮要揉得光滑，然後放進冰箱冰一段時間，以便於開酥時的包掟。其次在擀麵時用力要均勻，每次摺疊後要將麵團靜置一段時間讓麵筋充分鬆弛，便於操作，防止有破裂產生。最後要清掉多餘的擀麵用手粉，防止麵團表面結皮。

淮陽酥皮製作時要注意，包酥要均勻，收口不可太厚，擀酥要輕，厚薄均勻，防止皮乾，少用手粉，擀摺與包餡前要充分鬆弛。除此之外，掌握正確的爐溫是重點，過低的爐溫會令成品酥脆度不夠，過高會外焦內生，適當的爐溫在 180℃～ 220℃之間。

基礎酥皮步驟圖
Basic puff pastry

1 將水皮材料搓揉成光滑有彈性的麵團 ① ～ ② 。

2 將油皮材料搓揉出韌性或如圖所示的狀態 ③ ～ ④ 。

3 將水皮和油皮分別切成等份，將水皮捏出劑子（廣東人稱之為出蹄，因為形狀似馬蹄，意即從麵團上分出來的小塊）⑤ ～ ⑧ ，將油皮搓圓 ⑨ ～ ⑩ 。

4 把油皮包入水皮中，收緊收口 ⑪ ～ ⑮ ，按扁 ⑯ ～ ⑰ 。

5 擀成細長薄舌頭形 ⑱ ～ ⑳ ，再由上至下捲起 ㉑ ～ ㉒ 。

6 捲起後將麵團轉直輕壓 ㉓ ～ ㉕ ，再由左右向中心摺成三層 ㉖ ～ ㉜ 。

7 用擀麵棍擀開成薄圓形 ㉝ ～ ㉞ 。

1 Knead all water dough ingredients together until smooth and resilient ① ~ ② .

2 Knead all lard dough ingredients together until the mixture resemble mashed potato ③ ~ ④ .

3 Divide the water dough and lard dough into equal pieces separately. Roll them round. (water dough ⑤ ~ ⑧ , lard dough ⑨ ~ ⑩).

4 Wrap a ball of lard dough in a piece of water dough. Seal the seam ⑪ ~ ⑮ . Press flat ⑯ ~ ⑰ .

5 Roll it out into a long thin piece with a rolling pin ⑱ ~ ⑳ . Roll it up along the length ㉑ ~ ㉒ .

6 Press it flat again gently ㉓ ~ ㉕ . Fold in thirds from the sides toward the centre ㉖ ~ ㉜ .

7 Roll each dough out into a round disc ㉝ ~ ㉞ .

  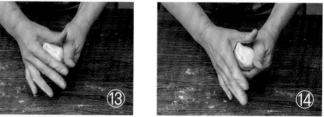

油皮包入水皮中。

包好，按扁。

擀成細長薄舌頭形。

捲起。

壓平,再擀薄。

再摺疊。

再擀開。

擀薄。

荷花酥

約可做
24 個

荷花酥，這道和荷花一樣清雅的中式酥點好玩又美麗。採用小包酥的做法，是浙江杭州著名的小吃。用油酥麵團製成的荷花酥，形似荷花，酥層清晰。觀之形美動人，食之酥鬆香甜，別有風味，是宴席上常用的一種花式中點，給人視覺上的享受。加上「出淤泥而不染」是人們對荷花高雅潔麗品質的讚譽，所以，高檔一點的素食店都做荷花酥給善信供奉觀音娘娘。

材料

水皮
低筋麵粉 .... 300 克
豬油 .... 95 克
水 .... 110 克
水溶性食用色素 .... 適量

油皮
低筋麵粉 .... 160 克
豬油 .... 65 克

餡料
白蓮蓉或豆沙 .... 360 克

炸油
花生油、蔬菜油或其他適合油炸的油 .... 適量

做法

1 將白蓮蓉或豆沙分為 24 份，每份約 15 克，搓圓。

2 將水皮材料按 P.40 酥點步驟圖 1 至 2 搓好，平均分成最少兩份，用色素調好 ① ～ ②，分成 48 份，出劑子，按基礎酥皮步驟圖 5 至 8，靜置 10 分鐘。油皮材料，按基礎酥皮步驟圖 3 至 4 搓好，分成 48 份，搓圓，按基礎酥皮步驟圖 9 至 10。

3 酥皮按照 P.40 - 43 基礎酥皮步驟圖 11 至 32。

4 按扁，擀成圓形，兩種不同顏色酥皮疊在一起，擀扁，每份分別包入一份餡料 ③ ～ ④。

5 於中央用小刀劃上米字花紋 ⑤ ～ ⑥，要劃至看到第二層酥皮，以不露餡為佳。

6 鍋內燒熱炸油至 130℃，將荷花酥分批放入鍋內 ⑦，炸約 5 分鐘至開花後，調高油溫約 140℃再炸約 1 分鐘，即可撈起瀝油。

Tips

1 不可一次炸太多，否則會讓油溫下降，荷花酥就難以慢慢開花。

2 油溫不可過高，適中的油溫才能讓酥慢慢開花，也讓酥餅變深色。

3 開花後提升油溫，可讓酥餅定型，不會散掉。

4 用小刀劃上米字花紋時，切記不要切到餡料，否則炸的時候，餡料會流出。

①

②

核桃酥

約可做
6個

用椰棗和法國核桃做成的核桃酥，過年用來送禮，
可說是一份既有心思又美味的禮物，收禮物的朋友一定捨不得吃！

材料

館料
椰棗 .... 50 克
烘香法國核桃 .... 60 克
砂糖 .... 13 克
糕粉 .... 6 克
水 .... 約 8 ～ 10 克

油皮
低筋麵粉 .... 55 克
固體豬油 .... 25 克

水皮
低筋麵粉 .... 65 克
豬油 .... 15 克
可可粉 .... 5 克
水 .... 30 克

做法

餡料

1 將椰棗、烘香核桃切碎，與其他材料混合，用少許力搓壓成圓形餡料如圖 ①～④。可按椰棗濕潤情況添加水分，稍為黏合便可，不用太濕。

皮料

1 製作酥皮可參閱 P40 ～ P43 的基礎酥皮步驟圖 1 至 34。

2 包入餡料 ⑤ ～ ⑥。

3 預留一點粉團作裝飾邊 ⑦，然後貼於中央 ⑧，以切刀在中央劃一刀 ⑨，用 11 號花鉗 ⑩ 劃上花紋 ⑪ ～ ⑫。

4 用 170℃的溫度，烤約 22 分鐘即可完成。

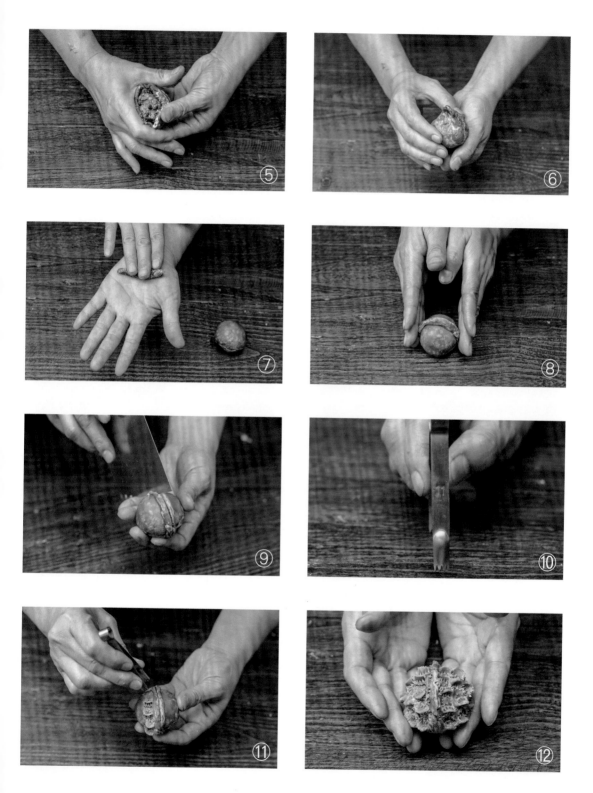

皮蛋酥

約可做
4 個

這是我執教的烹飪班上，最受歡迎的中式餅點，很多同學告訴我，他們的長輩都超喜歡。口味上，紅薑中和了皮蛋的鹼性和豬油的肥膩，偷偷告訴你們，開這課堂其實是出於我喜歡吃的私心啊！

材料

餡料

松花皮蛋	2 個
紅薑	100 克（每個 25 克）
白蓮蓉	200 克（每個 50 克）

油皮

低筋麵粉	80 克
固體豬油	40 克

水皮

低筋麵粉	90 克
豬油	20 克
水	40 克

做法

餡料

1 紅薑切碎。

2 松花皮蛋先將一個對切成兩份。

3 將蓮蓉擀開成薄圓形，將紅薑、半個皮蛋包入蓮蓉內，滾成圓形餡①～④。

皮料

1 將水皮材料搓揉成光滑有彈性的麵團⑤～⑦。

2 將油皮材料搓揉出韌性或如圖所示的狀態⑧。

3 將水皮和油皮分成 4 粒，搓圓，把油皮包入水皮中，收緊收口，擀成細長薄舌頭形，再由上至下捲起，捲起後將麵團轉直輕壓，再由左右向中心摺成三層⑨～㉘。

4 用擀麵棍擀開成薄圓形㉙～㉚。

5 包入餡㉛～㉞，塗上蛋黃㉟以220℃烤約 20～22 分鐘。

Tips

1 如果買到蛋黃較軟的皮蛋，過程中很難處理，可將其輕輕煮至蛋黃微微凝固(冷水下蛋，滾後慢火煮約 5 分鐘，再取出蛋過冷水)。

2 將蓮蓉包入紅薑皮蛋這動作不可過早處理，因為皮蛋鹼性會令蓮蓉出水，最好當天包當天做。

3 烤時間要掌握得好，如時間過長，蓮蓉會受熱膨脹，引起酥皮破裂。

將油皮包入水皮中。

擀成長薄狀。

捲起後輕壓。

再擀成薄長狀,由左右向中心摺。

左右向中心摺共三層。

再擀薄。

包餡。

整形,塗抹蛋汁,準備烘烤。

叉燒酥

約可做
12個

中學叛逆期時，常希望快點出社會工作、賺錢，還專挑時間長的工作做，就可以和家裡磨擦少一點。回想起來覺得很傻，大家年輕時總有過這念頭吧！記得第一份暑假打工，在餐廳當服務生，每天工作十小時，只有半小時的吃飯時間，因為剛開始工作的關係，常常見到美味的點心也很想吃，有時趁點心大佬不在時，就去偷他們的點心吃，他們叫「打貓」，有次被他捉個正著，偷的就是這叉燒酥。

材料

水皮

低筋麵粉	75 克
高筋麵粉	75 克
豬油	25 克
冰水	65 克
雞蛋	35 克

油皮

低筋麵粉	120 克
無鹽奶油	140 克
豬油	55 克
奶粉	20 克

餡料

叉燒	180 克
叉燒芡	130 克
洋蔥	半個 (切粒)

叉燒芡

砂糖	30 克
淡醬油	8 克
陳年醬油	8 克
太白粉	10 克
玉米粉	10 克
雞粉	2 克
海鹽	1 克
蠔油	15 克
胡椒粉、麻油	少許
清水	165 克
油	適量
蔥	數顆
洋蔥	1/3 個（切塊）
薑	2 片
紅蔥頭	2 粒

塗抹用

金黃糖膠	適量
炒香的芝麻	適量

做法

叉燒芡

1 用少部分的清水將太白粉、玉米粉開成勾芡水備用。

2 將砂糖、淡醬油、陳年醬油、雞粉、蠔油、胡椒粉、麻油、海鹽、水等混合。

3 起油鍋,爆香蔥、洋蔥、薑、紅蔥頭後 ① ,將步驟 2 倒入 ② 。

4 轉小火煮至出味,挑去薑蔥,倒入勾芡水拌勻成芡 ③ ～ ④ ,放涼備用。

餡料

1 將叉燒切碎,洋蔥炒香,拌入叉燒芡 ⑤ ,冷藏備用。

酥皮

1 水皮材料混合搓成軟滑麵團,放入冰箱冰約 30 分鐘。

2 油皮材料搓出韌性,滾成圓形,擀平,用保鮮膜包裹,放入冰箱冷藏約 30 分鐘 ⑥ 。

3 水皮擀平,水皮包入油皮 ⑦ ～ ⑧ ,用擀麵棍擀長 ⑨ ,將酥皮摺第一次 (三摺疊) ⑩ ～ ⑪ ,放入冰箱稍冷藏,擀長 ⑫ ,摺第二次(三摺疊),放入冰箱稍冷藏,擀長,摺第三次(四摺疊) ⑬ ～ ⑮ ,放入冰箱稍冷藏。

4 將酥皮擀開,用圓模壓出圓形 ,每份分別包入餡料 ⑰ ～ ⑱ ,封口向下,排在烤盤上,塗抹蛋黃 ⑲ ,放入 200℃ -220℃ 已預熱的烤爐內,烤約 18 分鐘至金黃色,用少許熱水調稀金黃糖膠,抹在叉燒酥表面,撒上炒香芝麻。

Tips

這叉燒酥採用大包酥的做法,三三四摺疊,麵團比較軟,要注意勿讓麵團黏在檯面,可把麵團放在厚帆布上製作。

擀長麵團。

第一次三摺疊。

擀長，再摺疊一次。

第三次摺疊。

壓出圓形酥皮,包入餡料。

封口朝下,排入烤盤,刷上蛋黃液。

花生酥

約可做
14個

這花生酥是由「京八件」演變而來的,所謂「京八件」就是八種形狀、口味不同的京味糕點,為清宮廷御膳房始創,流傳至民間。以棗泥、青梅、葡萄乾、玫瑰、豆沙、白糖、香蕉、椒鹽等八種原料為餡料,用油、水和麵做皮,以皮包餡,烘烤而成,取其色、香、味、形,都是造型精緻逼真、美觀兼味美的點心。

材料

餡料

花生	70 克
芝麻	20 克
松子仁	25 克
砂糖	40 克
海鹽	1 克
糕粉	8 克
豬油	8 克
花生醬	20 克
水	10 克

水皮

低筋麵粉	130 克
豬油	30 克
水	55 克
砂糖	15 克
即溶咖啡粉	3-4 克

油皮

低筋麵粉	100 克
豬油	50 克

做法

餡料

1 芝麻烘香，花生、松子仁切碎，和其他
材料混合成餡，分成 14 克一份 ①。

皮料做法

1 酥皮按 P.40 - 43 的基礎酥皮步驟圖 1
至 34。

2 包入餡料 ② ～ ③。

3 搓成小葫蘆型 ④，以花鉗由上而下鉗
上花紋 ⑤ ～ ⑦。

4 用 180℃烤約 15 分鐘。

Tips

1 也可用 espresso 即溶咖啡粉，它呈粉末狀較容易溶解，如用粒狀咖啡粉，
可以用份量內的水分先將其溶解。

2 建議使用 11 號花鉗鉗上花紋。

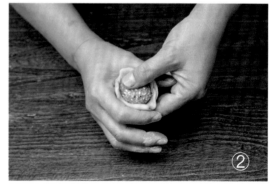

京蔥羊肉酥餅

約可做
10 個

這酥餅採用油酥麵團和膨鬆麵團,用小包酥的處理方法,把餅皮擀薄,做成皮薄餡多的美味酥餅。羊肉和京蔥更是絕配,一定能讓羊肉控欲罷不能。

材料

餡料

京蔥白	300 克
薑	少許
羊絞肉	350 克
鹽	6 克
砂糖	8 克
高湯	40-50 克
紹興酒	8 克
麻油	8 克
四川擔擔麵醬	30 克
孜然粉	6 克
太白粉	10 克

水皮

高筋麵粉	225 克
低筋麵粉	225 克
豬油	20 克
砂糖	10 克
水	250 克
鹽	2.5 克
新鮮酵母	8 克或速發酵母 3 克

油皮

低筋麵粉	225 克
豬油	105 克

裝飾

黑白芝麻	適量

做法

餡料

1 京蔥白、薑切丁,拌勻。

2 將羊絞肉和鹽用手摔打或用攪拌機攪至
出筋,一邊攪拌一邊加入高湯,然後加
入除了太白粉之外的調味料拌勻;加入
京蔥丁、薑和太白粉,拌勻成為肉餡
① ~ ⑤。

水皮、油皮

1 先做水皮。將水皮所有材料搓揉成光
滑有彈性的麵團,蓋上保鮮膜,發酵
45-60 分鐘,用手指沾麵粉插入麵團,
指印凹處不太會彈回來即可 ⑥。

2 將油皮的材料揉勻成泥狀。

3 接着把水、油皮各分成 10 等份,水皮
捏出劑子,油皮搓圓。

4 把油皮包入水皮中,收緊收口,擀成細
長薄舌頭形,再由上至下捲起。捲起後
輕壓,再由左右向中心摺成三層。(可
參考 P.40 - 43 的基礎酥皮步驟圖 11 至
32。)

5 用擀麵棍擀開成薄圓形,包入約 80 克
肉餡 ⑦ ~ ⑨,收口成圓球狀 ⑩,按
扁,表面噴水後沾上黑白芝麻 ⑪。

6 靜置約 10 分鐘後,再放入已預熱的烤
爐,以 180℃ -200℃烤約 25 分鐘即可。

芋蓉酥

約可做
15個

這芋蓉酥近似潮州月餅,以大包酥的方式製作。中秋節時,以這款
沒那麼甜膩的餅食和家人分享,老少咸宜!

材料

餡料

熟芋頭泥	270 克
玉米粉	25 克
奶粉	25 克
砂糖	70 克
椰漿	75 克
無鹽奶油	35 克
紫薯粉	20 克

水皮

低筋麵粉	200 克
砂糖	30 克
豬油	30 克
水	70 克
紫薯粉	20 克
水	5-10 克

油皮

低筋麵粉	180 克
豬油	100 克

做法

餡料

1 芋頭去皮、切片,蒸熟後馬上壓成芋泥 ①。

2 加入玉米粉、奶粉、砂糖、椰漿,拌勻,以大火蒸 15 分鐘後,拌入紫薯粉、奶油,搓勻成餡料 ②～④,分成等份。

水皮、油皮

1 將水皮材料(除了紫薯粉)搓揉至光滑且有彈性,將半份水皮加入紫薯粉和水 5-10 克搓成紫色麵團,靜置 10 分鐘。

2 將油皮材料搓至泥狀,分成兩份,搓成圓形 ⑤。

3 分別將兩色水皮各包入油皮 ⑥～⑦,用麵棒擀成長方形 ⑧ 再將酥皮摺第一次(三摺疊)⑨～⑩。

4 將酥皮擀成長方形,摺第二次(三摺疊),將酥皮擀成長方形。

5 將兩色酥皮疊起,中間灑水黏好 ⑪,四面切整齊,邊位壓薄,捲成約 6cm 直徑條狀 ⑫,分切約 1cm 厚等份 ⑬。

6 輕輕按扁,每份分別包入餡料 ⑭～⑰,封口向下,排在烤盤上,放入已預熱的烤箱內,以 170℃烤約 22 分鐘,但不需變金黃,即可取出待涼;或以 150℃ - 160℃油炸至熟透,注意油溫不可過高,否則會讓酥餅變深色。

Tips

1 按扁酥皮時要從螺旋紋處按下，這樣包入餡料後才美觀。

2 我個人較喜歡用炸的，愛其香脆口感。

胡椒餅

約可做
10 個

> 屬於發麵油酥類的淮陽酥皮胡椒餅，餅皮酥脆，餡料鹹香惹
> 味，加入白胡椒碎，更增添辣味，吃罷全身溫暖。

材料

水皮

高筋麵粉	225 克
低筋麵粉	225 克
豬油	20 克
砂糖	10 克
水	250 克
鹽	2.5 克
新鮮酵母	8 克

油皮

低筋麵粉	225 克
豬油	105 克

餡料

豬絞肉	500 克
砂糖	10 克
鹽	8 克
淡醬油	10 克
雞粉	5 克
清水或清雞湯	適量
紹興酒	少許
薑末	少許
麻油	少許
黑胡椒碎	15-20 克
白胡椒碎	5-10 克
葱末	375 克
太白粉	15 克

裝飾

白芝麻	適量

做法

1 將豬絞肉和鹽用手摔打或用攪拌機攪至起筋，一邊攪一邊加入清水或清雞湯，拌勻，再加入除了蔥末、太白粉之外的調味料拌勻①。

2 加入蔥末和太白粉拌勻成為肉餡②～③，冷藏備用。

3 先做水皮，將水皮所有材料搓揉成光滑有彈性的麵團，蓋上保鮮膜，發酵45-60分鐘，用手指沾麵粉插入麵團，指印凹處不太會彈回來即可。（可參考P.68京蔥羊肉酥餅的圖6）

4 將油皮的材料揉勻成泥狀。

5 接著把水、油皮各分成10等份，水皮捏出劑子，油皮搓圓。

6 把油皮包入水皮中，收緊收口，擀成細長薄舌頭形，再由上至下捲起。捲起後輕壓，再由左右向中心摺成三層。（可參考 P.40 - 43 的酥皮步驟圖 11 至 32。）

7 用擀麵棍擀開成薄圓形，包入約 50 克肉餡，收口成圓球狀，表面噴水後沾上白芝麻。

8 靜置約 10 分鐘後，再放入已預熱的烤箱，以 180℃ -200℃ 烤約 25 分鐘即可。

老婆餅

老婆餅

約可做
6個

在香港，以前想吃老婆餅一定要到元朗一帶，現在全港都有分店，可是和舊時的味道相比，總是差了點什麼。有些店已不再用冬瓜泥，改用什麼芝麻餡、抹茶餡等等，但我仍然覺得傳統的最好吃。我在這食譜內加入了桂花，讓餅添上一抹芳香。要吃好的，不如自己做吧！

材料

水皮

低筋麵粉	120 克
豬油	30 克
水	60 克

油皮

低筋麵粉	100 克
豬油	50 克

餡料

糖冬瓜	65 克
砂糖	45 克
糕粉	50 克
豬油	30 克
水	35 克
糖桂花	適量

做法

餡料

1 糖冬瓜用熱水浸泡，瀝掉水分讓甜味去除 ①。

2 用食物處理器將糖冬瓜打成泥後，和其他材料混合，搓揉成有彈性軟團 ② ～ ⑤。

3 搓好分成 6 份 ⑥。

皮料

1 將水皮材料搓揉成光滑有彈性的麵團。（可參考 P.40-43 的基礎酥皮步驟圖 1 至 2。）

2 將油皮材料搓揉成泥狀。（可參考 P.40 - 43 的酥皮步驟圖 3 至 4。）

3 將水皮和油皮分成 6 粒，搓圓，把油皮包入水皮中，收緊收口，擀成細長薄舌頭形，再由上至下捲起，捲起後輕壓，再由左右向中心摺成三層。(可參考 P.40 - 43 的酥皮步驟圖 9 至 32。)

4 用擀麵棍擀開成薄圓形。

5 包入餡，然後壓扁 ⑦ ～ ⑩，靜置 15 分鐘後，塗上蛋黃 ⑪，以 220℃烤約 15 分鐘。

包子、饅頭

包點屬於發酵類麵點，可依口味和製作方式分為：包餡與不包餡。包餡的有生煎包、壽桃等，不包餡的有饅頭、銀絲卷等。發酵麵點的發酵技術，經歷酒酵法、酸漿法、酵麵法、對鹼法和酵汁法等發展階段，至今仍然被採用。現在的發酵技術也更臻完善，其中包子和饅頭，更是發酵麵點文化的代表。以往包子和饅頭沒法分得清楚，後來以包餡的稱為包子，無餡的便稱饅頭。發酵麵點是以麵粉為主材料，加入不同水量和酵母或麵種，即可製作，再加入其他材料便可製作千變萬化的包點。

包點製作首先由麵胚開始，麵胚的原材料一般有麵粉、水、酵母、砂糖、油脂、鹽等組合而成。要做好包點，首先由選擇麵粉開始，麵粉種類繁多，如何選擇恰當的麵粉製作是關鍵所在。其次是水的比例，可用其他液體代替水，只要按比例加減便成。

以風味來說，廣東包點較鬆軟，口味亦較甜，會加入老麵種製作，一般會使用低筋麵粉，取其幼滑的口感，成品潔白細緻；北方人喜歡較有嚼勁的饅頭，他們會使用中筋麵粉甚至加入酵頭，讓饅頭更有風味。

養生饅頭

做饅頭過程中，大家會遇到許多問題，例如：塌陷、蒸出來的成品表面不夠光滑、水和麵粉的比例不對、大小不一、沒有彈性、口感不佳等等。要解決這些問題，首先由選擇麵粉開始，廣東人一般會使用低筋麵粉，取其幼滑的口感，做出來的成品潔白細緻。北方人喜歡較有嚼勁的饅頭，他們會使用中筋麵粉甚至加入酵頭，讓饅頭更有風味。

其次是麵粉和水的比例，一般的麵團比較乾身，這裡介紹的水分較適中，比較容易搓揉控制，可用其他液體代替水，只要按比例加減便成。學懂了製作白饅頭，便可以加點變化，配合使用天然食材，做出老少咸宜、健康養生，顏色亮麗的養生饅頭。

材料

椰汁牛奶饅頭材料（白色）

低筋麵粉（如喜歡有嚼勁的口感，可以換成 30 克高筋麵粉）.....................300 克
砂糖 ....35 - 40 克
發粉 .... 4 克
新鮮酵母 ....10 克
椰奶 (可以水或奶代替)................. 40 克
奶 .... 130 克

黃地瓜饅頭材料 (黃色)

低筋麵粉 .... 220 克
高筋麵粉 ....30 克
台灣黃地瓜粉 ....50 克
砂糖 ....35 - 40 克
發粉 .... 4 克
奶 ....175 克
新鮮酵母 ....10 克

紫地瓜饅頭材料 (紫色)

低筋麵粉 .... 220 克
高筋麵粉 .... 30 克
砂糖 ....35-40 克
發粉 .... 4 克
奶 .... 175 克
新鮮酵母 ....10 克
台灣紫地瓜粉 ....50 克

菠菜饅頭材料（綠色）

低筋麵粉 .... 300 克
冷凍菠菜泥（解凍，和奶打成菜泥）.......... 100 克
砂糖 .... 35 - 40 克
發粉 .... 4 克
新鮮酵母 .... 10 克
奶 .... 80 克

甜菜根饅頭材料（紅色）

低筋麵粉 .... 300 克
生甜菜根泥 .... 70 克
砂糖 .... 35 - 40 克
發粉 .... 4 克
新鮮酵母 .... 10 克
奶 .... 100 克

芝麻饅頭材料（花灰色）

低筋麵粉 .... 300 克
砂糖 .... 35 - 40 克
發粉 .... 4 克
新鮮酵母 .... 10 克
椰奶（可以水或奶代替）...................... 40 克
台灣黑芝麻粉或黑芝麻醬 .... 30 克
奶 .... 170 克

紅蘿蔔饅頭材料（橙色）

低筋麵粉 .... 300 克
生紅蘿蔔泥 .... 80 克
砂糖 .... 35 - 40 克
發粉 .... 4 克
新鮮酵母 .... 10 克
奶 .... 90 克

做法

1 乾材料量好後先放一半液體（蔬菜泥也一起放下）去，搓揉麵團使它成為雪花狀①，再慢慢加入其餘液體搓成硬身麵團②。

2 靜置 5-10 分鐘。如果沒有壓麵機，要將麵團搓滑一點，擀平，盡量去掉氣泡，重複，摺疊三～四次。如果有壓麵機③，麵團搓揉或攪拌的時間可減少，只要粗糙麵團便可以，因為還要將麵團過 3-4 次壓麵機，使麵團整齊平滑④～⑤。

3 將已壓好的麵團捲起來，捲時噴少許水在麵團上面以幫助貼緊麵團，捲緊成圓筒狀後，要考慮自己想造形的大小來搓長麵團⑥。切割時，刀要鋒利，分成⑦～⑧等份，不能切得太窄，否則很容易翻側。

4 蒸饅頭宜使用透氣好，不會讓水珠在蒸具內形成的竹蒸籠。饅頭造形後放在蒸籠發酵約 30-40 分鐘，每個饅頭之間要有適當距離讓它有空間發酵，不會黏在一起。天氣炎熱時，只需蓋上蓋子在室溫發酵；天氣冷時，用約 30-40℃的溫水搓麵，發酵時可以將蒸籠放在蒸鍋上，鍋裡放些熱水，以提高蒸籠裡的溫度。

5 蒸饅頭約 14 分鐘。蒸饅頭要用中火，如太小火，饅頭會不熟或蒸不透，吃來黏牙不爽口，鬆散沒彈性；大火又會爆裂或皺皮。蒸至中途要打開蓋，透氣一、兩次，或者不蓋緊鍋蓋，留一點點縫隙讓空氣流通。如果用金屬蒸籠，最好在蓋子包上毛巾吸收水蒸氣，以免滴下弄花饅頭表面。饅頭蒸好後不要立刻打開鍋蓋，等 2 分鐘才打開，讓其溫差不至於太大而使饅頭收縮。

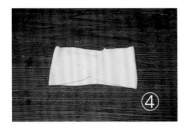  

Tips

1 麵皮靜置時間不宜過長，質感才會細緻
2 麵皮壓至適當厚度後，用手捲成圓柱體時是要搓緊的，而且動作要快，發酵才
  會均勻。
3 整形後要注意發酵時間，不足或過度都會對品質有影響。
4 蒸時要根據爐具火力而調節時間。
5 可以中筋麵粉全部代替低筋和高筋麵粉。

黃金鵝油
香蔥松子仁銀絲卷

約可做
16 個

偶爾在台北的永康商圈裡一間賣鵝油的專門店,看到了黃金鵝油香蔥,它揉合了法國鵝油和台灣紅蔥的精髓,可提升料理的美味。這道鵝油香蔥除了可伴飯、麵、青菜外,我還想到把它配上這清淡的銀絲卷,讓兩者發揮最佳的特色,味道鹹香雋永,相配得很。

材料

麵團

低筋麵粉	500 克
砂糖	45 克
發粉	7 克
新鮮酵母	18 克
水	140 克
奶	140 克

餡料

松子仁	100 克
海鹽	5 克
黃金鵝油香蔥	4-5 湯匙

做法

1　松子仁炒香或烘香，磨碎，和海鹽、黃金鵝油香蔥混合好候用 ① ～ ② 。

2　麵團做法按 P.86 養生饅頭做法步驟 1，將粉團分為兩份。

3　取一份麵團，平均切成兩塊，在每塊麵團上面塗滿黃金鵝油香蔥料 ③ 。

4　將兩塊麵團疊起 ④ ，麵團上塗滿黃金鵝油香蔥料，中間切開，再疊起 ⑤ ，平均切成條 ⑥ 。

5　將其餘黃金鵝油香蔥料，塗抹在切成條的麵團內。

6　另一份麵團切四份，擀長至可包入麵團條的長度，把麵團條包入 ⑦ 塗水在收口 ⑧ ，收口向下 ⑨ ，切成等份 ⑩ ，底黏烘焙紙，面放松子仁作裝飾，以中大火蒸約 14 分鐘。

Tips

也可以整條蒸好,再切開來吃。

壽桃包

約可做
42個

慶生的時候如果吃膩了蛋糕，可以改做這壽桃包來慶祝。我們家喜歡把一枚錢幣包在餡裡面，看誰吃了，便有紅包一枚，令氣氛更高漲、更溫馨，一定能逗老人家開心。

材料

皮料
低筋麵粉 .... 600 克
水 .... 280-300 克
新鮮酵母 .... 10 克
砂糖 .... 2 克

搓皮用料
砂糖 .... 40 克
固體豬油 .... 10 克
發粉 .... 5 克

餡料
白蓮蓉或豆沙 .... 450 克
水溶性食用桃紅色素 .... 少許

做法

1 將皮料搓成軟滑麵團，用保鮮膜包著，放在 25℃–32℃室溫發酵 45 - 60 分鐘，當發酵至約 2 倍大，可測試麵團。用食指沾少許麵粉，慢慢戳入麵團中，如麵團發酵適當，指孔就不會收縮 ①。

2 加入搓皮用材料，繼續搓至不黏手而麵團光滑，以保鮮膜包著醒發 20 分鐘。

3 將麵團分成每 25 克一份，揪成劑子，包入餡料，收口尾部搓長 ② ～ ⑥，墊上烘培紙，排入蒸籠 ⑦，最後發酵約 20 分鐘，蒸約 12 分鐘，表面按下時有回彈，便是熟透。

4 趁壽桃包還熱時，用竹籤或刀背在包身上壓上桃紋 ⑧。

5 用熱水調開水溶食用色素，用牙刷沾少許色水，在距離壽桃包約 6 – 8 吋距離，用小刀朝自己方向刮刷毛，讓色素水彈向壽桃包 ⑨。

Tips

如果壽包蒸熟後不夠尖，可用手指把頂端捏尖一點。

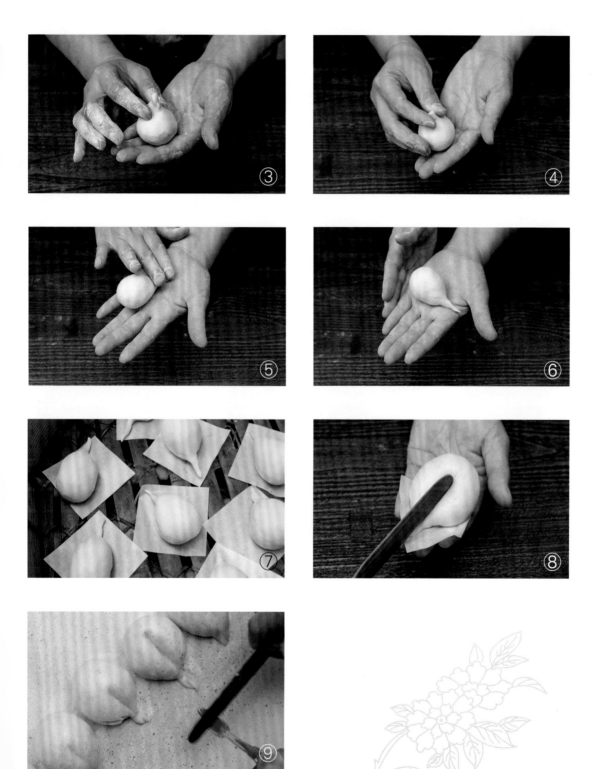

桂花棗饅頭

約可做
40個

軟甜的紅棗,清香的桂花,加上味道樸實的饅頭,
三者相得益彰。

材料

皮料

低筋麵粉	270 克
高筋麵粉	30 克
砂糖	30 克
發粉	5 克
新鮮酵母	10 克
水	120 克
鮮奶	40 克

餡料

雪棗或大紅棗	40 個
糖桂花	3 茶匙
冰糖	50 ～ 60 克
水適量	水適量

做法

1 將紅棗、冰糖和糖桂花放入鍋內,注入水,水要蓋過紅棗約 1cm,煮至紅棗稍軟(勿煮得太爛),讓糖水浸泡棗子至冷卻,瀝掉糖水(糖水可留下,另作他用)①。

2 皮料詳細做法,請參照 P.86 養生饅頭做法步驟 1。

3 將壓好的麵團捲起,搓長,切成 25 克一份。按扁,用壓麵機或擀成長條狀,用刀修邊 ②,包入一個紅棗 ③,露出 1/3,捏好底部 ④。

4 造型後,放在蒸籠發酵約 15 分鐘,每個之間要有適當距離讓饅頭有空間發酵,不會黏在一起 ⑤。

5 用中火蒸饅頭,約 6 分鐘,過程中需要打開蓋讓空氣流通一次。

煮軟紅棗,將麵團壓平厚切條。

包一顆紅棗,將底部收口。

先發酵 15 分鐘,再蒸 6 分鐘。

蝦醬生煎包

蝦醬生煎包

我最初接觸生煎包是在上中學年代，在冷冷的冬天，站在大鐵鍋前，心急地等著裡頭的生煎包出爐，沾滿豆瓣醬和鎮江醋，吃在口中，很是滿足，美味無窮。長大了，每次光顧上海飯店都喜愛叫這道小吃，可是總比不上街頭賣的好吃。有一次去上海，專程去找它的蹤影，在下榻的酒店旁的生煎包小攤，賣的包子地道、精緻又好吃。這款生煎包是我最愛的，包身既鬆軟又有嚼勁，皮厚薄適中，內餡肉嫩汁水豐富，包底煎得香脆，麵粉水底層又不會黏著鍋底。膨鬆的麵團加入了用冷水調沸水麵團的方法，讓麵團糊化，讓口感變得軟糯且帶有甜味，加上發酵麵團，讓口感鬆軟中帶有嚼勁，用上蝦醬調味更添上一份港式的地道風味。

材料

餡料

高麗菜	400 克
海鹽	少許
豬絞肉	600 克
蝦醬	30 克
砂糖	20 克
太白粉	8 克

燙麵麵團

中筋麵粉	400 克
海鹽	5 克
固體豬油	2.5 克
熱水（85℃ -100℃）	185 克
冷水（調節用）	5 克

發麵麵團

中筋麵粉	200 克
新鮮酵母	5 克
泡打粉	2.5 克
水	185 克
固體豬油	5 克

做法

餡料

1 高麗菜切絲後，用海鹽醃至軟，洗掉鹽
  分，擠乾水備用 ①。
2 豬絞肉用蝦醬、砂糖調味，放大碗內用
  手攪拌均勻。
3 加入高麗菜絲、太白粉拌勻成餡 ②。

燙麵麵團

1 麵粉、鹽、固體豬油略攪勻，加入熱水，
  用筷子迅速攪拌 ③，待蒸氣散去，用
  冷水調節麵團軟硬度，把麵團搓至光
  滑。

發麵麵團

1 把乾性材料略攪勻，加入水 (保留少許
  以調節濕度)。
2 搓成麵團後加入油搓滑。
3 把麵團滾圓，收口朝下，靜置 15 分鐘，
  此時製作燙麵麵團。

生煎包

1 發麵麵團與燙麵麵團搓勻 ④ ～ ⑤，滾圓收口，用保鮮膜包好，靜置 20 分鐘，即成生煎包麵團。

2 把麵團搓成長條，分成每 20 克一分 ⑥。

3 劑子壓扁，擀成周邊薄中間厚的圓片 ⑦，包入餡料 ⑧。

4 300 克冷水加入 20 克麵粉拌勻調成麵粉水。

5 平底鍋內倒入油，開小火熱鍋。

6 包子逐一排放入鍋中 ⑨，改中小火煎煮。

7 加入麵粉水，浸至煎包的 2/5 至 1/2 高度 ⑩，蓋上鍋蓋。

8 煎至水乾但皮還稍有黏性時，撒上蔥花、白芝麻 ⑪。

9 煎至底部金黃酥脆，淋上少許麻油。

10 離火等約 1-2 分鐘後，再小心地把包子倒扣在盤上 ⑫ ～ ⑬，以豆瓣醬、鎮江醋沾食。

④

⑤

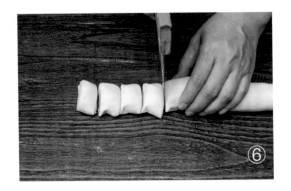
⑥

⑦

⑧

⑨

⑩

⑪

⑫

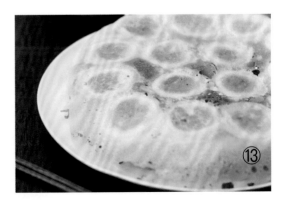

⑬

糕點

在中國，每個地方的糕點，各有不同的特色，除了糖果、糕餅、甜羹之外，還有各式各樣的風味鹹點。「糕」，是將麵粉、米粉或其他雜糧粉加入適量的液體，如水、蛋或油脂，調勻，蒸或烤而成。「點」，是小巧精緻，卻又有一定分量，味道好，在略感飢餓時候食用的小食品，比零食正式，但一般不能單獨地作為正餐。

糕點最大的特色，是那種充滿濃濃溫情的復古風格，所用的材料如：番薯、芋頭、蘿蔔、粉類、豆類等等，這些在街邊店舖裡司空見慣的日常食材，說不上有多麼特殊，也沒有特別創新，卻是歷久不衰的經典味道，會讓人有一種返璞歸真的感覺，也喚回小時候的點點回憶。

糕點的種類、製作方法繁多，不能盡錄，這章節中的食譜只能窺探一二。香港因地利因素，許多原材料也能進口，做糕點的變化也多，如中西食材並重的芝心湯圓、較新派的芒果糯米糍等等。好些我從小吃到大的糕點，也收錄在這書內，如缽仔糕、蘿蔔糕、芋頭糕等等。這些糕點易學難精，唯有不斷練習，才能掌握個中竅門。

冰皮月餅

約可做
25個

自有記憶以來,月餅就是我最愛的節慶食品。我最先會挑蓮蓉月餅
吃,之後是豆沙月餅,最不喜歡綠豆蓉和五仁。雖然這兩年奶黃月
餅成為亮點,亦有雪糕月餅、巧克力月餅,但冰皮月餅的市場,依
舊不可忽視。朋友鳳鳳教會了我做冰皮月餅,說比坊間的還好吃,
雖然我總覺得除了廣式月餅外,其餘的所謂月餅只是甜品,但這冰
皮月餅後一試難忘,鳳鳳果然說得沒錯。

材料

綠豆餡料

綠豆	225 克
乾桂花	4 克
無鹽奶油	180 克
鮮奶	480 克
砂糖	150 克
鹹蛋黃(蒸熟切碎)	6 個

皮料

低筋麵粉	25 克
在來米粉	50 克
糯米粉	60 克
砂糖	50 克
牛奶	250 克
煉奶	35 克
豬油	15 克
炒熟糯米粉(手粉)	適量

做法

綠豆餡

1　綠豆及乾桂花一起用水浸泡 4 小時（天氣炎熱時宜放入冰箱，以免餿掉）。

2　瀝掉乾桂花及水後，蒸熟綠豆。

3　綠豆蒸熟稍稍放涼，用攪拌機打成泥 ①。

4　無鹽奶油、綠豆泥同時倒入鍋內，炒熱；慢慢加入鮮奶、砂糖 ② ～ ③。

5　用慢火將綠豆泥炒至濃稠 ④，過篩瀝掉粗粒 ⑤，待涼後加入切碎的鹹蛋黃，每 40 克分成一份，滾圓，冷藏。

餅皮

1　豬油隔水加熱融化。

2　乾材料拌勻，少量地加入牛奶，搓至柔滑成團 ⑥ ～ ⑦。

3　加入全部牛奶後，再加入煉奶、豬油 ⑧。

4　所有材料混合後，倒入糕盤，大火蒸約 35 分鐘。

5　蒸熟後放涼，不要倒掉浮在面層的水及油 ⑨。

6　把皮料倒進大碗，搓揉至水油混合及有彈性 ⑩。

7　皮料每 20 克一份，包入冷藏的綠豆餡 40 克 ⑪ ～ ⑫。

8　月餅模撒上炒熟糯米粉，月餅團入模後，要用掌心用力壓實，才能使花紋明顯 ⑬ ～ ⑭，然後輕輕在檯面敲擊木模，讓月餅脫模，或以彈簧模印出月餅，立即放入冰箱保鮮 ⑮。

⑤

⑥

⑦

⑧

⑨

⑩

⑪

包入餡料後，搓圓放入模型。

用力壓下才能使花紋漂亮。

Tips

1 皮料蒸好後，揉搓、滾餡料和包餡時，都要戴上一次性使用的手套處理，以保衛生。

2 綠豆要分數次放入攪拌機內，以免把攪拌機弄壞。煮綠豆餡宜用不沾鍋，以免黏鍋。餡料成品軟硬度，應該像花生醬一般，冷藏後會再硬一點。

3 冰皮和餡料均可事先做好，放入冰箱可存放數天；做好的冰皮月餅放入冷凍，可存放 2 星期。

4 木月餅模皮餡共重 75 克，比例大約是一份皮料配二份餡料，本書使用的彈簧月餅模，比例則是皮 20 克、餡 40 克。

棗蓉糕

棗蓉糕

約可做 **1** 盤
20cm x 20cm
x 6cm 大小

紅棗可按大小分為大棗、小棗,按產區分為南棗和北棗,按乾濕分為鮮食和乾食。香港常用的不外是雞心棗、南棗。最近市面常見,如同果粒般巨大的新疆天山雪棗,它之所以稱為天山雪棗,原因之一是因為此棗用水質清涼無污染的天山冰雪融水灌溉。原本在新疆阿克蘇地區,是沒有種植大棗的,後來引進種植後發現,大棗味道甘甜,營養豐富,完全可與新疆哈密大棗和和田玉棗媲美。紅棗的香甜,我是很有規律地思念著的,尤其在女性月事後,身體總覺得虛弱,能喝著一碗暖烘烘的紅棗茶,溫暖自喉嚨透進丹田,頓覺身心舒泰。這道棗蓉糕是紅棗水的變奏,使用了兩種棗子,雙重補身,讚啊!

材料

新疆天山雪棗	300 克
紅棗	300 克
水	1000 克
片糖	150 克
泰國木薯粉	450 克
無筋麵粉	70 克

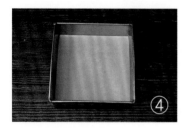

做法

1 紅棗加水蒸 3 小時 ①，蒸好後連水一起過篩，將紅棗壓成泥 ② 直到剩 1150 克紅棗水。

2 棗泥水加片糖煮至糖融化，糖煮融後，如果發覺水份量過多，須再開火煮一會讓水分蒸發，讓紅棗味更濃。

3 泰國木薯粉、無筋麵粉過篩，再慢慢加入已放涼的紅棗水中，拌勻至無粉粒 ③，再過篩。

4 將粉漿淨重量除以 8 份，以計算出每層應有的重量，在蒸每一層時倒入相同重量的粉漿，因為粉漿容易沉澱，所以不能一次倒入所有粉漿蒸熟，要逐層處理。

5 蒸盤塗油，倒入一層粉漿 ④（倒入前需將粉漿再拌勻一下），蒸約 4-5 分鐘至熟，重複至粉漿蒸完 ⑤。

6 紅棗糕放涼後，要存放在冰箱數小時至變硬，這樣才容易將紅棗糕切成工整的形狀，要食用時再蒸熱。

Tips

1 已蒸熟的紅棗連水在按壓時，不宜太用力，否則紅棗泥水會變得混濁。

2 可先嘗嘗紅棗泥水的甜度，再增減糖的份量。

3 採用新疆天山雪棗加雞心棗的效果更佳。

鮮椰汁馬豆糕

約可做 **1** 盤
20cm x 20cm
x 6cm 大小

馬豆糕並不是特別好吃，而是我對它的情結。我的二姨婆以前是在當富貴人住家工的，是馬姐（幫傭）一名，就像桃姐，煮得一手好菜。她來探望我們時總會帶著美味的椰汁馬豆糕。纔嘴的我總會忍不住偷吃，如不是老媽喝叱，一定吃個精光。所以我常常盼望她來探訪。

後來我請她教我做鮮椰汁馬豆糕，她說要用新鮮椰汁，雖然當時已出了罐頭椰汁，但她堅持用新鮮的。我張羅了很多地方都找不到，只能買一個鮮椰子回來自己用攪拌機攪汁。雖然不及椰子香料店賣的那麼香濃，但姨婆還算滿意，還讚我有誠意。當然她是沒有食譜的，我一邊看，一邊將材料磅秤，整理出食譜，也在她面前再試做，於是她的古法馬豆糕就學成了。在市面上能吃到的，大多是加了奶，讓糕身顏色潔白，有的加了魚膠或洋菜凍，好讓口感清爽，但都不是我喜歡的口感。姨婆做的只用玉米粉，竅門是用新鮮椰汁，因有椰油，做出來的糕才會爽滑，也要向同一方向攪拌才行。我試過用罐頭椰汁，做出來的糕不是太粉便是不清爽，差強人意。有時候，老人家堅持的東西，總是有她的道理的。

材料

馬豆（黃豌豆）.. 100 克
鮮椰汁（或自製鮮椰子汁 1900 克，就可免去
下方列出的水分）.. 600 克
水 .... 1300 克
玉米粉 .... 200 克
砂糖 .... 280 克

做法

1 先將馬豆用水煮至軟，但不煮爛 ①
（左邊是未煮馬豆，右邊是已煮的）。

2 把一半水與玉米粉拌勻備用 ②。

3 其餘的水與椰汁、砂糖煮至冒煙。

4 一邊慢慢倒入玉米粉水 ③，一邊朝
同一個方向攪拌至成稀糊和滾起，小
心煮焦。

5 下馬豆 ④，煮至成稀糊及滾起 ⑤，
熄火。

6 倒入模中 ⑥ ～ ⑧，放涼後，放入冰
箱冷藏。

自製鮮椰子汁

1 一個椰子挖起肉、刮去咖啡色皮，約
600 克，椰子水 600 克。

2 加水的份量不要過多，大約 1 杯椰肉
用 1.5 杯水攪拌，先攪出椰汁，濾渣
後，再混合椰子水即可。

Tips

1 把椰子的咖啡色皮刨去，椰汁會更潔白。

2 鮮椰子汁在賣椰子的香料店或印尼雜貨店有售，鮮椰汁很易變壞，購買後要盡
　快放入冰箱保鮮；椰汁的脂肪含量較高，需要濾掉多餘脂肪才製作糕點，以免
　糕點過於油膩。

3 如買不到鮮椰子汁，可用椰子肉加水攪拌，加入椰子水，味道會更濃郁；攪拌
　後切記要濾掉椰子泥才成汁。

4 玉米粉的用量也很重要，放多了糕點太扎實，並要慢慢攪拌至粉漿熟透，不然
　吃起來會變得粉粉的。

5 用了鮮椰汁就不需要放奶，因為椰汁有天然的白色。

五彩番薯
南瓜糕

約可做 **1** 盤
23cm x 23cm
x 6cm 大小

大部分女生都愛澱粉類的食物，應該會喜歡這道。在這款糕
點我用了不同顏色的番薯，還加了芋頭和椰汁，過年時帶回
公司，連男同事都愛死它。

材料

玉米粉	50 克
在來米粉	300 克
熟南瓜泥	600 克
清水（視南瓜水分）	200-300 克
椰漿	250 克
砂糖	150 克
生紫心番薯	200 克
生黃心番薯	200 克
生橙心番薯	200 克
生芋頭	200 克

做法

1 南瓜連皮切大塊，去核，隔水蒸熟，把南瓜肉挖出壓成泥。

2 番薯、芋頭切丁。

3 熟南瓜泥、砂糖和 1/3 分的水，放入鍋內煮滾。

4 在來米粉、玉米粉用椰漿和清水調勻。

5 將煮滾的南瓜泥糖水加入粉漿內 ①，立即攪拌均勻。

6 加入番薯、芋頭丁攪拌均勻 ② ～ ③。

7 放入已塗油的糕盤內 ④ 蒸約 1.5 小時。

8 冷卻後切片，蒸熱食用。

Tips

食譜所用的水分視乎南瓜的含水量，中國南瓜的水分高，可以用食譜刊中的最少水量開漿；如使用日本南瓜，因肉質較粉，水分少，以食譜中的最多水量開漿。

芒果糯米糍

芒果糯米糍

約可做 **2** 條
20cm x 5cm
x 4cm 大小

芒果糯米糍，其實不是什麼民間小吃。我在一間酒樓吃過這種做法後，我私心地把心愛的芒果布丁食譜融入其中，這食譜可分作兩道小吃。皮料做好後，可包不同的餡料，例如：榴槤、芒果肉、草莓或花生芝麻餡。

材料

餡料

芒果泥 ....150 克
椰奶 .... 35 克
鮮奶油 ....35 克
砂糖 ....50 克
滾水 ....75 克
冰水 ....75 克
吉利丁 ....9 克
芒果肉 ....適量

皮料

糯米粉 .... 200 克
玉米粉 .... 40 克
砂糖 ....100 克

奶粉 .... 40 克
罐頭椰汁 .... 250 克
奶 ....160 克
水 .... 200 克
油 .... 40 克

表面覆蓋用料

椰蓉 ....適量

做法

餡料

1 將芒果去皮取出果肉後，用食物處理器或攪拌機攪成果泥，過篩，瀝去雜質。

2 吉利丁片用冷水泡軟。

3 滾水加入砂糖煮溶後，再加入已泡軟的吉利丁片，輕輕攪拌使其溶化。

4 將椰奶、鮮奶油、冰水倒入吉利丁糖水中，加入芒果泥，攪拌均勻 ① ～ ② 。

5 倒入沖過清水的模具中，再放入切成大塊的芒果肉，放入冰箱冷藏至凝固 ③ 。

皮料

1 將粉料放在盆子內，加入砂糖、奶、罐頭椰汁、清水調勻，加入油攪勻 ④ ～ ⑤ ，過篩，瀝去雜質。

2 倒入方形淺盤內，用大火蒸 20-25 分鐘至熟 (視乎厚度，蒸到脹起便熟)，成糯米皮 ⑥ ，放涼備用。

組合

1 將芒果餡切成所需大小。戴上手套，拿一團已涼的皮料，沾少許椰蓉當手粉，壓扁成長方形，包入一條芒果布丁 ⑦ ，捲好 ⑧ ，滾上椰蓉，修好邊角 ⑨ ～ ⑩ ，放入冰箱冷藏即可享用。

將芒果布丁包入皮料中。

修好邊角，沾上椰蓉，冷藏。

Tips

1 如製作冰皮月餅般，捲製糯米糍時要戴上一次性使用手套處理，以保衛生。

2 皮料一定要放涼後再包上餡料，否則餡料會融掉。

3 要等糯米糍凝固後才能切片。

缽仔糕

缽仔糕

家中收藏了很多缽仔糕的食譜，試了很多都不太對味，不是太粉就是太硬。聽人説某店舖賣的缽仔糕好吃，我也會去買來試，結果都是差強人意。

獨角仙住的社區，每天下午 2 點左右，有一對伯伯和嬸嬸，推著手推車，載著剛做好的缽仔糕由白沙灣到西貢墟擺賣，他的缽仔糕是我吃過最好吃的，爽滑清甜，蔗糖味十足。無論在他剛開始營業或即將收攤時碰到他，我會立即買來吃，遇到他要收攤的時候，他總給我很多缽仔糕。我幾度想衝動地叫他傳授這民間美食的秘技，了卻心頭之念，可惜 伯伯已不在。

獨角仙鑽研食譜是很死心眼的，一是不做，要做就是一百次、一千次，也要做到滿意為止，於是找來食譜，做了十幾二十次，最滿意這個，也最近似伯伯做的。記著要等缽仔糕稍涼才吃，吃時才爽口，熱熱吃太糊了。

材料

熟紅豆粒	適量
蔗片糖	1 片
在來米粉	100 克
無筋麵粉	22 克
荸薺粉	10 克
水	500 克
蔗片糖（切碎）	120 克

做法

1 生紅豆洗淨，泡發，加水和 1 片蔗片糖，煮至軟，但不要攪拌太頻繁，以免弄破紅豆。最後，用水沖去紅豆沙。

2 用 140 克的水把在來米粉、無筋麵粉和荸薺粉調勻。

3 用 360 克的水，混合已切碎的片糖，煮至完全溶解。

4 糖水煮滾放入粉漿內 ①，一面放入一面攪勻 ②，加入部份紅豆粒 ③。

5 把稀漿倒入已塗油的小碗中 ④，再放入適量紅豆粒在糕面 ⑤ ～ ⑥，最後隔水蒸約 15-20 分鐘 (視糕的大小而調整時間)。

6 蒸熟後稍為冷卻，用竹籤把缽仔糕取出。

Tips

不同牌子的無筋麵粉的軟硬度都不同，我在香港用寶山牌澄麵粉粉，做出來的糕點較爽口。你可以試著找出最喜歡的品牌來使用。

健康雜糧糕

約可做 **1** 盤
23cm x 23cm x 6cm 大小

有一次去廣州從化泡溫泉，早餐時在酒店吃到這道雜糧糕，清淡可口。請教了師傅做法，回來試了十來次不同的份量，以這食譜最對胃，減肥的我們可以多吃點囉！

材料

糕料

荸薺粉	220 克
木薯粉	85 克
水（A）	330 克
砂糖	155 克
水（B）	725 克

雜糧

紅豆	120 克
鷹嘴豆	220 克
小米	60 克
玉米粒	100 克
砂糖	適量
水	適量

做法

1 分別將紅豆和鷹嘴豆浸泡 4-8 小時 (天氣炎熱時宜放入冰箱，以免餿掉)。

2 浸泡好的紅豆，用水煮約 30 分鐘，倒掉水分去除苦澀，再加水和適量砂糖，煮軟後，瀝掉水，備用。

3 將已浸泡 4-8 小時的鷹嘴豆沖洗乾淨，加適量水和砂糖煮至軟，瀝掉水，備用。

4 分別將小米、玉米粒沖洗乾淨，加適量水煮至軟，瀝掉水，沖水去除膠質，用手壓掉水分，備用 ① 。

5 荸薺粉和木薯粉用水（A）調勻，濾掉雜質。

6 水（B）和砂糖煮滾，將荸薺粉漿放入糖水內，攪拌均勻，粉漿應該變成稀糊狀，以用匙羹舀起再倒下，不會有紋理的狀態為佳 ② 。

7 預備蒸盤，在盤內塗油。

8 將粉漿和雜糧調勻，粉漿份量以能鋪平蒸盤表面為合適 ③ 。

9 最下層倒入小米粉漿，填滿整個盤底 ④ ，大火蒸約 7 分鐘 ⑤ 。

10 取出，倒入紅豆粉漿 ⑥ ～ ⑦ ，大火蒸約 10 分鐘。

11 取出，倒入鷹嘴豆粉漿，大火蒸約 10 分鐘。

12 取出，倒入玉米粒粉漿，大火蒸約 7 分鐘。

13 蒸好後放冰箱冷藏，再切塊食用。

①

②

③

Tips

1 倒入粉漿前要攪拌均勻，以免粉漿沉澱。

2 蒸熟後的雜糧糕要冰硬了才切，會更美觀，線條更美。

3 雜糧可隨意轉變，只要注意按雜糧大小，估算煮軟時間。

4 搭配不同顏色的雜糧，糕點會更美。

5 雜糧可多煮，過水後可放入冷凍存放幾個月，要用時拿出來解凍便可。

芝心湯圓

約可做
8粒

這道芝心湯圓的靈感來自電視節目。那天被那名廚師的仔細俐落的手法和新穎的食物吸引著，看見主持人指著這款芝心湯圓直說好吃的樣子真情流露，決定做來試試，一試難忘。

材料

餡料

奶油乾酪 (或鹹味較輕的奶油乾酪) 80 克

砂糖 ....40 克

烘香花生碎 ....40 克

烘香芝麻 .... 少許

皮料

糯米粉 ....100 克

無筋麵粉 ....10 克

砂糖 ....20 克

熱水 ....80 克

固體豬油 ....12 克

表面用料

椰蓉 ....150 克

砂糖 ....40 克

烘香花生碎 ....80 克

做法

1 將全部餡料攪勻，稍稍冷藏，每份 18 克，分成 8 份 ① ～ ② 。

2 將全部皮料搓成粉團，每份 25 克，分成 8 份。

3 全部表面用料拌勻；表面用料的份量可隨意添減。

4 用一份搓好的皮料包一份餡料 ③ ～ ⑤ ；搓圓，用滾水將湯丸煮約 8 分鐘或看見芝心湯圓浮起，即可用隔篩撈起，立即放在表面用料上 ⑥ ，滾上一層表面用料，即可趁熱食用。

134

椰汁年糕

約可做 **1** 條
23cm x 23cm
x 6cm 大小

這款年糕在我家已經出現了二十多年，記得當年在裕華國貨見到椰糖，感覺椰味十足，便拿來試做年糕，一做便年年都做，比起坊間的椰汁年糕更椰味十足，不相信？試做看看吧！

材料

印尼椰糖 ....400 克
清水 ....250 克
糯米粉 ....530 克
無筋麵粉 ....80 克
椰漿 ....450 克

做法

1 先將椰糖用清水煮溶，放涼備用。

2 糯米粉、無筋麵粉用椰漿開好後，慢慢
  地加入糖水 ①，一面倒入一面攪勻，
  切勿有粉粒，過篩 ②。

3 倒入已塗油蒸盤內 ③～④，按蒸盤大
  小蒸至年糕變深色，用竹籤插入，如沒
  有淺色粉漿黏著，代表已蒸熟。

Tips

無筋麵粉可使糕點爽口，如想要軟糯的口感，也可以不放。

魚湯鯪魚肉
蝦乾蘿蔔糕

約可做 **1** 盤
26cm x 26cm
x 6cm 大小

猶記起以前的房東太太，她是順德人，擅長烹調淡水魚菜式，過年的時候，她常做這味魚湯鯪魚肉蝦乾蘿蔔糕，很客氣地請我吃。這蘿蔔糕清甜無比，她用了自己熬製的魚湯代替清水，完全沒有放味精，沒臘味的肥膩，多吃了也不會造成負擔。

材料

蝦乾	80 克
鯪魚肉（已調味）	500 克
油	少許
在來米粉	450 克
胡椒粉、麻油	少許
清水	1200 克
白蘿蔔	2000 克
芫茜（香菜）、葱（切粒）	適量
魚湯寶（魚高湯塊）	2 粒
玉米粉	70 克
海鹽	25 克

做法

1 蝦乾切粒，白蘿蔔去皮刨成細絲狀。

2 鯪魚肉、芫茜、葱混合好。

3 用少許油將鯪魚肉煎成鯪魚餅 ①，將
  鯪魚餅切小塊。

4 蘿蔔絲、蝦乾用 2/3 分量的清水煮軟
  後，加入魚湯寶和切小塊鯪魚餅再煮
  滾 ②。

5 在來米粉、玉米粉、胡椒粉、麻油、
  海鹽用其餘清水調勻。

6 將粉漿倒入已煮滾的蘿蔔湯內 ③，
  立即攪拌均勻 ④ ～ ⑤。

7 倒入已塗油的蒸盤內 ⑥，蒸約 1.5 小
  時。

8 冷卻後將蘿蔔糕切成丁狀，並煎至金
  黃。

Tips

不同牌子的在來米粉有不同的軟硬度，配合玉米粉使用時，可隨喜歡的口感調整用量。

客家鹹年糕

約可做 **1** 盤
26cm x 26cm
x 6cm 大小

鹹年糕是台灣苗栗的客家傳統年糕，古法是用糯米和在來米泡水後，磨成米漿，脫水後成粉膏，再搓揉成團做年糕，很費功夫，所以過年才能吃到。材料中的紅蔥頭也要手切和爆香才有風味，最後用手搓勻材料和麵團才正宗。現在有現售的糯米粉和在來米粉，方便得多，但處理材料的功夫就不能省了。

材料

冬菇（中型）	6 朵
乾蝦米	80 克
梅花肉（略帶肥肉）	300 克
紅蔥頭片	180 克
油	適量
頭抽	兩茶匙
砂糖	75 克
糯米粉	600 克
在來米粉	300 克
胡椒粉、麻油	少許
海鹽	25 克
雞粉	8 克
清水	800 克

做法

1 冬菇泡軟切丁，蝦米泡軟洗淨，梅花肉切大片，用份量外的鹽、糖、玉米粉略醃，紅蔥頭切片。

2 紅蔥頭片用油爆香至金黃 ①，依序放下蝦米、冬菇丁爆香，最後放梅花肉 ②、頭抽爆香。

3 在來米粉、糯米粉、胡椒粉、麻油、砂糖、海鹽、雞粉混合，倒入清水搓軟成團 ③。

4 粉團混合已爆香材料 ④，搓勻 ⑤ ～ ⑥。

5 倒入已塗油的蒸盤內蒸約 1.5 小時。

6 冷卻後切片，煎至金黃食用。

臘味芋頭糕

臘味芋頭糕

約可做
5斤

做芋頭糕最重要的竅門是芋頭一定要夠粉，粉糯的芋頭能夠吸收臘腸、臘肉的甘香，再搭配提味的五香粉，真是非常美味。選購芋頭時，拿起來時要輕，如果重重的，表示芋頭水分多，過程中會有「出水」的情形。

材料

臘腸 ....80 克
臘肉 ....80 克
蝦米 ....15 克
芋頭 .... 450 克
水 ....1750 克
在來米粉 ....338 克
玉米粉 ....20 克
紅蔥頭 .... 2 粒
干貝 ....15 克

調味料

鹽 .... 20 克
砂糖 ....38 克
雞粉 ....10 克
五香粉 .... 9 克
胡椒粉、麻油 ....各少許
油 .... 適量

做法

1 將臘腸、臘肉出水後切丁，蝦米、干貝泡軟備用。

2 將芋頭切丁，過油 ①，再用 2/3 清水煮滾 ②。

3 用大碗將玉米粉、在來米粉混合餘下的清水和調味料，調勻備用。

4 起油鍋爆香紅蔥頭，盛出紅蔥頭，再爆香臘味、干貝、蝦米 (預留少許臘味不爆香，留待最後鋪上表面)。

5 芋頭煮軟後加入臘味，煮滾 ③ ～ ④。

6 粉漿攪勻，將粉漿倒入已煮軟的芋頭中 ⑤，攪勻 ⑥，倒進塗了油的糕盤內，蒸約 90 分鐘至熟透，糕面放預留臘味。

7 放涼後，芋頭糕放入冰箱冷藏，再切片煎香享用。

①

②

③

④

⑤

⑥

Tips

1 切芋頭時,可切些小丁的使其煮散成糊,這樣會更香、更美味。

2 五香粉的份量可按喜好加減。此外,一般街市售賣的包裝五香粉,味道比超市
買的瓶裝五香粉更佳。

3 倒入粉漿時水要熱,否則粉漿不夠濃稠,芋頭粒會浮起。

家鄉蒸鬆糕

約可做 **1** 盤
直徑 26cm 大小

家鄉蒸鬆糕是一年才有機會吃一次的高點，夾著煎堆（芝麻球）吃是很有風味的，是老媽家鄉的賀年食品。外婆健在時曾請她教我，可惜未有機會實習過，後來請了好幾次九姨婆（婆婆的第九妹妹）教我，總是時間配合不上。

有一次終於可以見到她，我立刻拿了相機把她的説話，原原本本地錄下，免得食譜失傳。一開始攝錄，原來我老媽也懂得做，一邊攝錄九姨婆、她一邊答嘴，哎，真搞笑。好啦，原來每年九姨婆都是從麵包店買糕種（麵種 A）做糕。再經過我的研究後，試用淨粘米粉加酵母來發種，不成功呀。其後我再想，一定不會是酒樓的老麵種吧！因為她在麵包店買的發子一定是酵母，第二次就想通了，用了一半的粘米粉加一半的麵粉再發一次，成功了，連 5 元也省了。雖然這糕不是特別好吃，但蘊含著很多的記憶和情意。

材料

麵種 A 材料

在來米粉 ....100 克
低筋麵粉 ....100 克
新鮮酵母 ....10 克
水 ....120 克
砂糖 ....20 克

麵種 B 材料

在來米粉 ....600 克
麵種 A ....全部
麵粉 ....100 克
菜油 ....200 克
水 ....300 - 400 克

糖水材料

片糖 ....600 克
水 ....400 克

做法

1 將全部麵種 A 材料搓好，放在溫暖地方（26℃-28℃）發酵約 7-10 小時 ①～②。

2 加入全部麵種 B 材料，搓滑放在溫暖地方（25℃-28℃）發酵約 4 小時 ③。

3 用水將片糖煮溶放涼，過篩備用。

4 將已發麵種慢慢加入糖水 ④，用手拌成稀漿 ⑤～⑥，瀝掉雜質，再放於溫暖處（32℃-35℃）發酵約 4-5 小時，至起泡 ⑦，攪拌均勻 ⑧，倒進已塗油的蒸盤，蒸約 1.5 小時。

Tips

老媽說每次、每鍋只可蒸一盤，否則會不漂亮；同時蒸的時間也要掌握好，因為再蒸的是不會熟的。

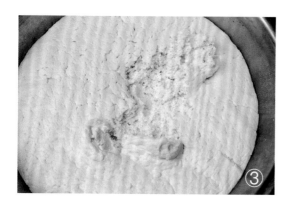

③

④

⑤

⑥

⑦

⑧

Understanding Chinese buns and snacks

Tens of Chinese buns and snacks are covered in this cookbook and I included the methods and special tricks for the best results. From rustic homely dumplings to fluffy crispy puff pastry, I covered traditional buns as well as innovative cakes and sweets. Before you tread into my realm of cakes and snacks, let's familiarize ourselves with the common ingredients used, types of dough and their methods, and how to make different fillings.

Common ingredients used in Chinese buns and snacks

Flour

Flour can be divided into 4 main categories according to their protein content high gluten flour, bread flour, all-purpose flour and cake flour.

High gluten flour, or super strong flour, contains at least 15% of protein. It picks up lots of water and makes highly resilient dough. It works great in noodles.

Bread flour has protein content between 11.5 and 14.5%. It makes elastic dough and is suitable for making pan-fried pork buns, spring onion flatbread, and wonton or dumpling skin.

All-purpose flour has protein content between 9 and 10%. It picks up medium amount of water and makes dough of medium elasticity. It is used in buns.

Cake flour contains the least protein ranging between 7.5 and 8.5%. It has a fine texture and makes the end product look snow white. It is commonly used in Cantonese buns, knife-sliced plain buns, steamed cake etc. Common brands include American Roses and Narcissus from Hong Kong.

Glutinous rice flour

It is ground short-grainglutinous rice. It makes dough that is soft, sticky and chewy after cooked. It is used to make Tangyuan (glutinous rice dumplings in soup) and Nian Gao (New Year cake).

Long-grain rice flour

As its name suggests, it is ground long-

grain rice and it is used in many different food items. It has the lowest elasticity among all ricegrains and it is used in steamed Chinese cakes with a loose starchy texture, such as radish cake or taro cake.

Starches

Starch is another key component in cakes and snacks. It can come from different plants, such asgrains, tubers and vegetables. The most common ones include wheat starch, cornstarch, sweet potato starch, tapioca starch, potato starch, and water chestnut starch.

Wheat starch is actually bread flour that has been rinsed, rubbed and precipitate to remove all gluten. Without any gluten, wheat starch makes dough that is crisp in texture, without any chewiness. It is also highly formable before cooked and looks translucent after cooked. It is commonly used in Chinese dumpling skin (e.g. Hargow, the shrimp dumpling, and Fenguo, the Chaozhou dumpling) and other Chinese cakes.

Cornstarch, potato starch, tapioca starch and water chestnut starch are mostly used in making the filling, as a coagulant or a thickening agent. You may also use them in conjunction with other starches to change the properties of the dough.

Grease

Oil andgrease improve the texture of dough so that the end product is softer and moister in texture. It is also used in filling. In fried food, it is the heat conducting agent. Commonly usedgrease and oil include lard, butter and vegetable oil.

Sugar

It adds sweetness to food, achieves balance of palate,gives the end product a certain colour and keeps the end product soft while extending its shelf life. Commonly used sugar includes white sugar, icing sugar,golden syrup, light brown sugar, dark brown sugar and rock sugar.

Egg

Eggs are very versatile in cake and pastry making. Adding eggs to filling or deep-frying batter gives an eggy aroma. It also helps the food brown nicer while making the texture finer and more elastic. Eggs also have nutritional value. Commonly used eggs include fresh eggs, salted eggs and thousand-year eggs.

Yeast

It is a live leavening agent with much rising power. It makes the end product fluffy with airy crumbs and texture.

Baking powder

It is a chemical leavening agent that introduce gas bubbles in the food. The gas expands when heated to make the food fluffy and airy.

Water

It is used to regulate the consistency and stiffness of dough, batter and filling.

Salt

It makes food salty and enhances the flavour of the ingredients. It makes a batter or dough brown more slowly.

Common types of dough in Chinese cake and buns

To make a dough, starches, such as flour, rice flour or other starches, are mixed with water,grease and eggs. It is then kneaded repeatedly to incorporate well and cooked in different ways (such as steamed, boiled, seared, pan-fried, deep-fried or baked) to make various forms of snacks and staples (such as noodles, congee, cake, crackers, rolls, buns, dumplings, thick soup and aspic). In order to make the perfect Dim Sum, it's important to master the skills of dough making.

There are 5 types of dough used in Chinese Dim Sum

Wheat starch dough

Boiling hot water is poured into wheat starch or a wheat starch slurry is cooked to make the dough. Wheat starch dough is snow white in colour with a translucent look when rolled out thin. The texture is fine and soft, making the end product silky smooth to the palate. It is used in multigrain cake, coconut pudding with yellow split peas or Put Chai Ko.

Puff pastry dough

It is made with a lard dough and a water dough. Sometimes eggs, sugar and chemical leavening agents are also added. It can be further classified into layered puff, cookie puff, and deep-fried puff. Layered puff pastry have distinct layers of crumbs and it can be open (layers visible from outside) or closed (layers invisible from outside). Layered puff pastry is crispy in texture and lovely in colour. It is used to make barbecue pork pastry, thousand- year egg pastry, walnut pastry and taro pastry.

Leavened dough

Yeast or chemical leavening agent is added to the dough so that airy structure is created within the crumb. The end-product tends to have light and spongy texture, plump and full outline. It is also nutritious and its nutrients can be absorbed by human readily. Examples made with this dough include steamed plain buns, longevity peach buns and silver thread rolls.

Water dough

It is unleavened dough made with water and flour. The water can be cold, hot or warm. It is used in all sorts of buns and snacks including pan-fried pork buns.

Rice flour dough

It is a batter made with rice flour, water, seasoning and oil. It can be shaped before steamed or steamed before shaped. Examples include radish cake and homestyle dumplings.

Making the filling

The filling can be savoury or sweet in taste. It can be vegetarian or meat-based. It can also be pre-cooked or raw when wrapped in the dough.

Any dry ingredients used in the filling should either be rehydrated in water or steamed until soft first before used. If the ingredients carry unwanted scent or taste, such as gamey smell, bitterness or acridness, you should remove that scent or taste first. Ingredients with much moisture content (such as vegetables) should be marinated first to draw out most of its moisture.

The consistency and stickiness of the filling will directly affect the quality of end product and how you shape the buns or dumplings. When you make vegetarian filling, try to dry the ingredients so that the filling can be bound more tightly. When you make a filling with raw meat, add some water or pork skin aspic to bind the ingredients. Cooked meat doesn't stick well and you may thicken the filling with a glaze. Adding grease, egg, or thick sauces also helps increase the stickiness. When you mix the filling ingredients, try to do it quickly and evenly. Otherwise, moisture may be drawn out of the ingredients and the filling may become watery.

Finally, pay attention to how finely or coarsely the ingredients are diced. The seasoning and the consistency of the filling should also complement the taste and texture of the dough.

Homestyle dumplings

Homestyle dumplings, or Cha Guo in Chinese, are typical Hakkanese snacks. They come in many varieties and can be savoury or sweet in taste. Savoury filling includes radish, black-eyed beans, dried shrimps and ground pork. Sweet filling includes bean paste, peanuts, sesames, sweet potato, pumpkin or skunkvine. There are also plain ones without filling, such as Xi Ban and Cha Ban, both of which are chewy and cake-like. At one time, homestyle dumplings were very difficult to find because they are complicated to make. Young people are not interested in learning and only very few grandmas were making them and selling them. In recent years, activists are keen to resurrect local culture and small businesses. People are once again interested in authentic old-time Hong Kong food and Cha Guo becomes a representative snack appealing to hikers and visitors to the New Territories and offshore islands. Obviously, it is a lot more readily available than before and there are many new flavours on top of the traditional ones.

I personally have strong feelings for these rustic snacks accessible to the masses. It's because they taste great, of course. But I'm even more fascinated by the process of making them, which was always a team effort across three generations. The whole family came together to make Cha Guo as offerings to deities before festivals. It's such a memorable and heart-warming scene to see daughters, mothers andgrandmas working side by side for these dumplings.

My grandma used Java olive leaves as liners for her dumplings because she had a Java olive tree in her backyard. When the season was right, there were fiery-red pods hanging from the branches. Java olive fruit is very starchy and sweet, sort of like chestnuts. It tastes great after blanched in water or stewed with chicken.

But eating Java olive fruit actually needs quite a bit of work. You must remove the red pods to extract the brown nut. Then, remove the layers of soft skin and shells on the nut with your hands or a paring knife to expose the bright yellow nut. Boil them in lightly salted water until the shell cracks open and dig in.

Nowadays, Java olive leaves are hard to come by. You may use banana leaves or bamboo leaves instead.

Homestyle pumpkin dumplings with red bean filling

Red bean filling

cooked red beans (the cooking method see steps 1 to 3) 300g

water

sugar 100g

lard 30g

Dumpling skin

Japanese pumpkin (steamed till soft) 450g

glutinous rice flour 40g

long-grain rice flour 60g

brown sugar 240g

lard 15g

water (for regulating dampness)

Dumpling liners

Java olive leaves, banana leaves or bamboo leaves

Red bean filling

1 Soak the raw red beans until soft. Put them into a pot. Add water and cook over high heat for 30 minutes. Drain. This step helps remove the bitter taste of the red beans.

2 Put the beans back in the same pot. Add water and cook over high heat until the beans are starting to soften. Drain.

3 Add water again. Cook over low heat and boil until all red beans are split open. Drain again.

4 Put the beans into a non-stick pan. Add sugar and lard ① . Stir over low heat until all moisture is absorbed and the mixture turns into a thick paste.

5 If you prefer the filling with finer texture, you may mash the beans while stirring them. Let cool and use as a filling.

Method

1 Blanch the dumpling liners in boiling water briefly. Cut into desired shapes.

2 In a large bowl, put inglutinous rice flour, long-grain rice flour, brown sugar and lard.

3 Steam the pumpkin till cooked and mash finely. Put the mashed pumpkin to the glutinous rice flour mixture while still hot. Stir with a wooden spatula and knead into dough ② . (Amount of water added depends on water content of pumpkin.)

4 Divide the dough into 50-g pieces. Wrap 30g of red bean filling in each piece ③ . Brush oil on each piece of leaf and put a dumpling on top. Brush oil over the dumplings again.

5 Steam over high heat for 13 minutes.

Tips

1 Make sure you pay attention to the water content of the pumpkin. Chinese pumpkin tends to be more watery than Japanese ones. Make sure you adjust the amount of water added according to the dampness of the dough.

2 From September to November every year, Japanese pumpkins are in season and they are less steeply priced. Yet, make sure you read the labels carefully. Some pumpkins are "Japanese varieties" that aren't actually grown in Japan. They couldn't quite compare in terms of taste and quality. I once tried Japanese pumpkins grown in Mexico and New Zealand and they tasted okay. You may want to avoid those from other origins.

3 Leave your pumpkins on the counter for a few days before using. They'd taste even sweeter as they ripen.

Homestyle dumplings with black-eyed bean filling

Serves: makes 16 dumplings

Filling
dried black-eyed beans 300g
water
dried shrimps 50g
cooking oil 2tbsp
finely chopped shallot

Dumpling skin
glutinous rice flour 320g
long-grain rice flour 55g
sea salt 30g
lard 2g
warm water 300g

Seasoning
sea salt 8g
sugar 10g
five-spice powderground 5g
ground white pepper
sesame oil

Dumpling liners
Java olive leaves, banana leaves or bamboo leaves

Garnish

water-soluble food colouring

Filling

1 Boil black-eyed beans in water for 40 minutes until soft.

2 Rinse the dried shrimps. Soak in water until soft. Finely chop them.

3 Heat oil in a wok. Stir-fry shallot and dried shrimps until fragrant. Add black-eyed beans and seasoning. Mix well ① .

Method

1 Blanch the dumpling liners in boiling water briefly. Cut into desired shapes.

2 Mix glutinous and long-grain rice flour with salt.

3 Add warm water and lard. Knead into dough.

4 Divide the dough into 50-g pieces. Wrap some filling in each piece. Seal the seam ② . Roll it round and put it on a piece of greased leaf ③ .

5 Brush oil on the dumplings. Steam for 13 minutes.

6 Make a dot on each dumpling with food colouring ④ .

Tips

Do not overcook the black-eyed beans. For the best result, the beans should be in whole but very tender after cooked.

Homestyle beetroot dumplings with sweet potato filling

Sweet potato filling
hot mashed sweet potato 250g
sugar 20g
unsalted butter 30g

Dumpling skin
cooked sweet potato 160g
cooked beetroot 160g
glutinous rice flour 220g
brown or dark brown sugar 120g
water 50g
lard 10g

Dumpling liners
Java olive leaves, banana leaves or
bamboo leaves

Method
1 To make the filling, mix all
   ingredients and stir well.

2 Blanch the dumpling liners in boiling water briefly. Cut into desired shapes.

3 To make dumpling skin, steam raw sweet potato and raw beetroot with skin on until done.

4 Peel the sweet potato and press through a fine wire mesh to mash it. Peel beetroot and grate it.

5 Cook brown or dark brown sugar in water until it dissolves. Pour the glutinous rice flour, cooked sweet potato and cooked beetroot into the syrup while hot. Stir with a wooden rolling pin until well mixed ① .

6 Knead until well incorporated. Add lard and knead again ② .

7 Take a piece of dough from step 6 (the size depends on the size of your mould). You may wrap some filling with the dough or leave it plain without filling ③ . Dust the mould withglutinous rice flour. Tap it to remove excess flour. Press the dumpling into the mould. Unmould it and put it over agreased leaf ④ .

8 Steam over high heat for 12 minutes.

Tips

1 Your dumplings may vary in colour depending how much pigment the beetroot contains.

2 Steam the dumplings of different sizes in separate batches. Do not steam for too long either. Otherwise, the dumplings may over-expand and the embossed pattern on them may warp or disfigure. They won't look as good that way.

3 You may also shape the dumplings into round patties without using the moulds.

4 Do not pour in all syrup at once. Adjust the amount used according to the water content of the sweet potato and beetroot.

5 Use ginger-flavoured brown sugar for more complex flavours.

Homestyle radish dumplings

Serves: makes 12 dumplings

Filling

white radish 300g

shallot

Chinese preserved pork sausage 45g

dried shrimps 15g

dried radish 15g

spring onion

Dumpling skin

warm water ..... 310 to 315g

glutinous rice flour 330g

long-grain rice flour 90g

lard 20g

sea salt 3g

Seasoning

sugar 6g

chicken bouillon powder 2g

sea salt

ground white pepper

sesame oil

oil

caltrop starch slurry

lard 15g

Dumpling liners

Java olive leaves, banana leaves or bamboo leaves

Filling

1 grate the radish into thick strips. Soak the dried shrimps till soft. Dice the shallot, pork sausage and dried radish.

2 Blanch the radish strips in boiling water until soft. Drain.

3 Heat oil in a wok. Stir fry shallot, pork sausage, dried shrimps and dried radish until fragrant. Put in the radish strips and seasoning.Stir until the radish picks up the seasoning.

4 Stir in caltrop starch slurry and add lard. Mix well. Add diced springonion and mix well. Let cool ① .

Method

1 Blanch the dumpling liners in boiling water briefly. Cut into desired shapes.

2 Sieve the glutinous and long-grain rice flour together. Add warm water and knead into a smooth dough that doesn't stick to your hands ② . Use more or less water according to the consistency of the dough.

3 Take 60g of dough and roll into a ball with an indentation at the centre. Put in 35g of filling and pinch to seal the seam well ③ . Put it onto a greased leaf with the seam side down. Press to flatten it into a round patty ④ .

4 Steam over high heat for 15 minutes. Serve.

Tips

1 You may alter the proportion between-glutinous and long-grain rice flour according to your preferred texture. The more glutinous rice flour you use, the stickier the dumpling is.

2 Traditionally, Hakkanese press a wooden sphere into the dough to make a bowl of out of it. I used a marble globe instead for the same purpose. I got the marble globe from Cat Street market.

Homestyle skunkvine dumplings

Serves: makes 27 dumplings

Filling
shelled peanuts 80g
black sesames 10g
white sesames 15g
brown sugar 40g

Dumpling skin
skunkvine leaves 200g
water 500g
raw cane sugar slab 80g
glutinous rice flour 500g
long-grain rice flour 60g
lard 30g

Dumpling liners
Java olive leaves, banana leaves or bamboo leaves

Filling

1 Rinse the sesames.

2 Toast peanuts and sesames until fragrant. Grind finely. Add brown sugar and mix well.

Method

1 Blanch the dumpling liners in boiling water briefly. Cut into desired shapes.

2 To make the dumpling skin, rinse the skunkvine leaves. Put them into a mortar and add some water. Pound with a pestle or blend the mixture in a blender. Transfer into a pot (with the solid bits). Add raw cane sugar slab and cook until sugar dissolves. Let cool slightly ① .

3 Put all dry ingredients into a mixing bowl. Add skunkvine syrup from step 2 a little at a time. Knead into dough. Adjust the amount of syrup used according to the dampness and consistency of the dough ② .

4 Take 50g of dough. Wrap in some filling. Pinch the seam to seal well. Put it on a greased dumpling liner with the seam side down ③ . Press gently into a round patty. Brush oil over it ④ .

5 Steam over high heat for 15 minutes. Serve.

Tips

1 You can find skunkvine in the rural area of Hong Kong easily. You can alsoget it from fresh herbal stores. Yet, skunkvine leaves come in two varieties – velvety hairy ones and smooth shiny ones. The Hakkas only use those with smooth shiny surface to make dumplings.

2 In the old days, Hakkanese tended to pick whatever available around them as liners for homestyle dumplings, such as banana leaves, bamboo leaves or Java olive leaves. I used Java olive leaves this time because of their generous sizes and shiny surface. They can stand prolonged steaming without breaking down and they add a lovely fragrance to the dumplings without being too overwhelming.

Chinese
puff pastries

Traditional Chinese puff pastry comes in two categories – Huaiyang style and Cantonese style. Typical Huaiyang puff pastries include Xie Ke Huang, chrysanthemum pastry with bean filling, puff pastry pouch topped with sesames, peppery pork buns and puff pastry with radish filling. Typical Cantonese puff pastries includes various kinds of wedding pastries (e.g. red lotus seed paste pastry, white five-nut pastry, yellow mung bean paste pastry, thousand-year egg pastry) and wife cake.

There are two ways to make Chinese puff pastry – the big wrap and the small wrap. The big wrap is similar to making Western puff pastry. A big piece of lard dough is wrapped with a big piece of water dough before folding and rolling out. The big wrap is efficient and fast in making a big batch of puff pastry dough. However, it's not easily to make layers of even thickness. There are also fewer layers and the pastry won't puff up much. In a home kitchen, the small wrap is usually preferred. The lard dough and water dough are divided into individual portions before wrapping, folding and rolling out. As the dough is smaller in size, it's a lot easier to handle. The puff pastry tends to have more layers and puff up more. Yet, it takes time to wrap, fold and roll out each dough for each pastry, so that it doesn't work for bulk production.

Cantonese puff pastry dough is made with lard, flour, eggs and sugar. After series of folding and rolling out, the puff pastry is fluffy and light, with beautiful colour and rich flavours.

Huaiyang-style puff pastry can be classified into the following types according to the method of folding

• Open layers and folds visible from the outside

• Closed layers can't be seen from outside; only visible after cut open

• Semi-closed layers and folds visible on part of the pastry

• Fermented puff pastry lard dough is wrapped in yeasted dough before folding and rolling out

Both Chinese and Western puff pastries work on the same principles. Solid grease is wrapped in dough and forms hundreds of thin layers after repeated folding and rolling out. The grease separates the flour layers. In the baking process, the air in between the flour layers heats up and expands. That's why the pastry puffs up while giving exceptional crispiness and layered crumbs.

Cantonese puff pastry requires some skills in regulating the dough and rolling out the dough. The water dough needs to be kneaded till smooth. It is then rested in a fridge for a while so that the dough yields better when shaped and divided. When your roll the dough out, make sure you apply the same force all over. After each fold, let the dough sit briefly to rest it. Otherwise, it might lose its elasticity and crack. Lastly, brush off any excess flour on the surface as it tends to form a hard crust after a while.

To make Huaiyang-style puff pastry, make sure the layers are even in thickness. When you fold or wrap it, try not to make the seam too thick. Roll out the dough gently so that it is of even thickness all over. To prevent drying out the dough, sprinkle as little flour on it as possible before rolling out. Rest the dough thoroughly before rolling, folding or wrapping in filling.

Last but not least, baking temperature is also a key factor. Pastry baked at a temperature too low tends to be not crispy enough. Those baked at a temperature too high tends to be burnt on the outside while the core is still uncooked. The preferred oven temperature for Chinese puff pastry ranges between 180°C and 220°C.

Lotus pastry

Water dough

cake flour 300g

lard 95g

water 110g

food colouring

Lard dough

cake flour 160g

lard 65g

Filling

white lotus seed paste or red bean paste 360g

Oil for frying

peanut oil, vegetable oil or corn oil

Crust, assembly and baking

1 Divide the filling into 24 equal portions, each weighing 15g. Roll each portion into a ball.

2 To make the water dough, follow steps 1 and 2 of "Basic puff pastry" recipe on p.40. Divide into at least 2 equal parts. Add food colouring to one part and knead the two pieces of dough separately ① . Divide each dough into 48 equal parts. Then follow steps 5 to 8 of "Basic puff pastry" recipe on p.40. Let them sit for 10 minutes. To make the lard dough, follow steps 3 and 4 of "Basic puff pastry" on p.40. Divide into 48 equal pieces. Roll them round and follow steps 9 to 10 of "Basic puff pastry" recipe on p.41.

3 Then wrap the lard dough in the plain water dough. For wrapping, rolling and folding, follow steps 11 to 32 of "Basic puff pastry" recipe on p.41-43.

4 Press the pastry dough flat. Roll it into a round disc. Stack the coloured water dough on the pastry dough ② . Roll them flat. Wrap in one piece of filling.

5 Make light cuts on the pastry with a sharp knife in the form of an eight-spoked asterick. Try to cut through the pastry dough only without cutting through the coloured water dough ③ .

6 Heat oil in a wok up to 130°C. Put in the pastries in a few batches. Deep fry for 5 minutes until the crust opens up ④ . Turn the heat up to 140°C and fry for 1 more minute. Drain. Serve.

Tips

1 Do not deep fry too many pastries at one time. Otherwise, the oil temperature would drop too quickly and the pastries would pick up too much oil. The crust also won't open up properly.

2 Do not fry the pastry in oil too hot. Keep the oil at optimal temperature so that the pastry crust has time to slowly pop open while gaining a golden colour.

3 After the crust opens up, fry the pastry over higher temperature to firm up the "petals" so that they are less likely to fall apart.

①

②

③

④

Walnut pastry

Filling
palm dates 50g

toasted French walnuts 60g

sugar 13g

fried glutinous rice flour 6g

water ..... 8 to 10g

Water dough
cake flour 65g

lard 15g

cocoa powder 5g

water 30g

Lard dough
cake flour 55g

lard 25g

Filling

Finely chop the palm dates and toasted walnuts ①. Mix well other filling ingredients ② Roll and press the mixture into a round patty. Add just enough water to bind the filling. It should not be too wet.

Crust, assembly and baking

1 Follow steps 1 to 34 in "Basic puff pastry" recipe on p.40-43.

2 Divide the dough into 6 pieces. Roll flat and wrap filling into thedough.

3 Set aside some dough for decorative trim ③ . Put the trim at the centre.

4 Make a cut at the centre with a knife ④ . Make patterns on the cut with a pastry crimper ⑤ .

5 Bake in a preheated oven at 170°C for 22 minutes. Serve.

Thousand-year egg pastry

Filling

thousand-year eggs 2

red candiedginger (about 25g for each pastry) 100g

white lotus seed paste (about 50g for each pastry) 200g

Water dough

cake flour 90g

lard 40g

water 20g

Lard dough

cake flour 80g

lard 40g

Filling

1 Finely chop the candied ginger.

2 Slice each thousand-year egg into halves along the length.

3 Roll out each piece of white lotus seed paste into a disc. Wrap in 25g of candied ginger and half of a thousand-year egg. Roll it into a ball.

Crust, assembly and baking

1 Knead all water dough ingredients together until smooth and resilient ① .

2 Knead all lard dough ingredients together until the mixture resemble mashed potato ② .

3 Divide the water dough and lard dough into 4 equal pieces separately. Roll them round. Wrap a ball of lard dough in a piece of water dough. Seal the seam and roll it out into a long thin piece with a rolling pin. Roll it up along the length. Press it flat again gently ③ . Fold in thirds from the sides toward the centre.

4 Roll each dough out into a round disc ④ .

5 Wrap in the ball of filling. Brush on some egg yolk on top. Bake in a preheated oven at 220°C for 20 to 22 minutes. Serve.

* For the crust, follow steps 1-34 in " Basic puff pastry" recipe on p.40-43.

Tips

1 The best thousand-year eggs tend to have runny yolks. Before you use it in the filling, you may cook them slightly so that the yolks are less runny. First, remove the rice hulls on the thousand-year eggs. Put them in a pot of cold water. Bring to the boil over low heat and cook for 5 minutes after the water boils. Rinse in cold water. Then shell and use in the filling.

2 After you wrap the candied ginger and thousand-year egg in the lotus seed paste, do not let it stand for too long. The alkaline in the thousand-year egg tends to draw moisture out of the lotus seed paste. Make sure you wrap the filling in pastry within the same day.

3 When you bake Chinese flaky pastry, make sure you control the baking time precisely. The lotus seed paste may expand too much if over-baked and the pastries may crack.

Barbecue pork pastry

Serves: makes 12 pastries

Water dough
cake flour 75g
bread flour 75g
lard 25g
ice water 65g
eggs 35g

Lard dough
cake flour 120g
unsalted butter 140g
lard 55g
milk powder 20g

Barbecueglaze
sugar 30g
light soy sauce 8g
dark soy sauce 8g
caltrop starch 10g
cornstarch 10g
chicken bouillon powder 2g
sea salt 1g
oyster sauce 15g
ground white pepper
sesame oil

water 165g

oil

spring onion

onion (wedged) 1/3

ginger ..... 2 slices

shallots 2

Filling

barbecue pork 180g

barbecueglaze 130g

onion (diced) 1/2

Garnish

golden syrup toasted sesames

Barbecue glaze

1 Mix caltrop starch and cornstarch with part of the water to make a slurry.

2 Mix sugar, light and dark soy sauce, chicken bouillon powder, oyster sauce,ground white pepper, sesame oil and salt together.

3 Heat a wok and add oil. Stir fry spring onion, onion,ginger and shallot until fragrant.Add the seasoning mixture from step 2.

4 Cook over low heat until the flavours mingle well. Remove theginger and spring onion. Stir in the slurry from step 1 and cook until it thickens ① . Let cool.

Filling

Finely chop the barbecue pork. Stir fry onion in oil until fragrant. Stir in the barbecue glaze. Refrigerate for later use ② .

Crust, assembly and baking

1 Mix all water dough ingredients together and knead into smooth dough. Refrigerate for 30 minutes.

2 Mix all lard dough ingredients together until it resembles mashed potato. Roll into a ball and wrap in cling film. Refrigerate for 30 minutes.

3 Roll out the water dough with a rolling pin. Wrap the lard dough inside the water dough ③ . Roll it out into a long piece. Fold the dough in thirds for the first time. Refrigerate briefly. Roll it out into a long piece again. Fold in thirds for the second time. Refrigerate briefly and roll it out into a long piece. Fold in quarters for the third time. Refrigerate briefly.

4 Roll out the dough and cut out round discs with a cookie cutter. Wrap some filling with each piece of dough. Put them on a baking tray with the seam facing down ④ . Brush some egg yolk on top. Bake in a preheated oven at 200°C to 220°C for 18 minutes until golden. Thin out the golden syrup with some hot water. Brush on the pastries. Sprinkle with toasted sesames. Serve.

Tips

This thirds-thirds-quarters pastry folding method is call the "big wrap." The dough tends to be softer in texture and make sure you don't roll it directly on the counter. It's easier to handle on thick canvas.

Ground peanut and sesame pastry

Filling
peanuts 70g
sesames 20g
pine nuts 25g
sugar 40g
sea salt 1g
fried glutinous rice flour 8g
lard 8g
peanut butter 20g
water 10g

Water dough
cake flour 130g
lard 30g
water 55g
sugar 15g
instant coffee 3 to 4g

Lard dough
cake flour 100g
lard 50g

Filling

Toast peanuts, sesames and pine nuts separately until lightly browned. Finely chop peanuts and pine nuts. Mix all filling ingredients together. Divide into 14-g pieces ①.

Crust, assembly and baking

1 Make the dough according to steps 1 to 34 in the "Basic puff pastry" recipe on p.40-43.
2 Wrap in a piece of filling ②.
3 Roll into a calabash shape ③. Crimp patterns on each pastry with a pastry crimper ④.
4 Bake in a preheated oven at 180°C for 15 minutes. Serve.

Tips

I prefer espresso instant coffee powder in the water dough because it tends to dissolve more readily. If you use instant coffee granules, you may want to dissolve them in the water first before mixing in with other ingredients.

Lamb and Peking scallion pastry

Filling

Peking scallion (white parts only) 300g

ginger

ground lamb 350g

salt 6g

sugar 8g

stock 40-50g

Shaoxing wine 8g

sesame oil 8g

Sichuan Dandan noodles sauce 30g

ground cumin 6g

caltrop starch 10g

Water dough

bread flour 225g

cake flour 225g

lard 20g

sugar 10g

water 250g

salt 2.5g

fresh yeast .. 8g or instant yeast .. 3g

Lard dough

cake flour 225g

lard 105g

Garnish

black and white sesames

Filling

1 Dice the Peking scallions and ginger. Mix well.

2 Stir ground lamb with salt until sticky. Pour in stock slowly while stirring continuously. Add the remaining seasoning and stir well. Put in diced Peking scallion,ginger and caltrop starch. Stir and set aside ① .

Crust, assembly and baking

1 To make the water dough, put in all ingredients and knead into smooth and resilient dough. Cover in cling film. Let it rise for 45 to 60 minutes. Check if the dough has risen enough by poking it with your finger. If the indentation remains, it has risen enough ② .

2 To make the lard dough, mix all ingredients until it resembles mashed potato.

3 Divide the water dough into 10 equal pieces. Divide the lard dough into 10 equal pieces. Roll the lard dough into balls.

4 Wrap one piece of lard dough inside one piece of water dough. Seal the seam and roll it out into a long piece. Roll it up along the length.

5 Press the dough flatgently. Fold in thirds from the sides towards the centre.

6 Roll it out into a round disc. Wrap in about 80g of filling ③ . Seal the seam and roll it round. Press gently. Spray water on top and sprinkle with black and white sesames ④ .

7 Let the pastries sit for 10 minutes. Bake in a preheated oven at 180°C to 200°C for 25 minutes. Serve.

* For the crust, follow steps 1-34 in "Basic puff pastry" recipe on p.40-43.

Mashed taro pastry

Serves: makes 15 dumplings

Filling
cooked taro 270g
cornstarch 25g
milk powder 25g
sugar 70g
coconut milk 75g
unsalted butter 35g
purple sweet potato powder 20g

Water dough
cake flour 200g
sugar 30g
lard 30g
water 70g
purple sweet potato powder 20g
water ..... 5 to 10g

Lard dough
cake flour 180g
lard 100g

Filling

1 Peel and slice the taro. Steam until soft. Mash finely while still hot.

2 Stir in cornstarch, milk powder, sugar and coconut milk. Mix well and steam for 15 minutes over high heat. Stir in purple sweet potato powder and butter ① . Knead into filling. Divide into 15 equal parts.

Crust, assembly and baking

1 Mix all water dough ingredients (except purple sweet potato powder) and knead until shiny and resilient. Divide the dough into two parts. Add purple sweet potato powder and 5 to 10g of water to one part of the dough. Knead until well incorporated. Let the dough sit for 10 minutes.

2 Mix all lard dough ingredients together to achieve mashed potato consistency. Divide into two parts. Roll each part into a ball.

3 Wrap one part of lard dough in the plain water dough. Wrap the other part of lard dough in the purple water dough ② . Roll each out into a rectangle. Fold each dough in thirds.

4 Roll the dough out into a long rectangle with a rolling pin. Fold it in thirds for the second time. Roll it into a rectangle again. Repeat with the other dough.

5 Stack the one piece of dough on the other. Smear some water in between to secure. Trim the four edges neatly. Roll the edges thin and roll into a log about 6cm in diameter ③ . Cut into 1-cm thick slices.

6 Press each piece of dough flat. Wrap in some filling ④ . Put the pastries on a baking tray with the seam facing down. Bake in a preheated oven at 170°C for 22 minutes. Remove from oven before the crust turns golden. Let cool and serve. Alternatively, deep-fry the pastries in oil at 150°C to 160°C until cooked through. Make sure you keep the oil at the specified temperature. Otherwise, the pastries may turn out too browned.

Tips

1 When you press the sliced dough flat, follow the spiral as you press. The pastries will look better after fried that way.

2 I personally prefer deep-frying the pastries instead of baking, because of their crispy mouthfeel.

Peppery pork buns

> *Serves: makes 10 pastries*

Water dough
bread flour 225g
cake flour 225g
lard 20g
sugar 10g
water 250g
salt 2.5g
fresh yeast 8g

Lard dough
cake flour 225g
lard 105g

Filling
ground pork 500g
sugar 10g
salt 8g
light soy sauce 10g
chicken bouillon powder water (or chicken stock) 5g

water (or chicken stock)

Shaoxing wine

dicedginger

sesame oil

ground black pepper 15-20g

ground white pepper 5-10g

diced spring onion 375g

caltrop starch 15g

Garnish

white sesames

Filling

1 Add salt to the ground pork and stir until sticky. Pour in water or chicken stock slowly while stirring continuously. Add the remaining seasoning and stir well ① .

2 Add diced spring onion and caltrop starch. Mix again ② ~ ③ . Refrigerate for later use.

Crust, assembly and baking

1 To make the water dough, put in all ingredients and knead into smooth and resilient dough. Cover in cling film. Let it rise for 45 to 60 minutes. Check if the dough has risen enough by poking it with your finger. If the indentation remains, it has risen enough.

2 To make the lard dough, mix all ingredients until it resembles mashed potato.

3 Divide the water dough into 10 equal pieces. Divide the lard dough into 10 equalpieces. Roll the lard dough into balls.

4 Wrap one piece of lard dough inside one piece of water dough. Seal the seam and roll it out into a long piece. Roll it up along the length. Press the dough flat gently.Fold in thirds from the sides towards the centre.

5 Roll it out into a round disc. Wrap in about 50g of filling. Seal the seam and roll it round. Press gently. Spray water on top and sprinkle with white sesames.

6 Let the pastries sit for 10 minutes. Bake in a preheated oven at 180°C to 200°C for 25 minutes. Serve.

* For the crust, follow steps 1-34 in "Basic puff pastry" recipe on p.40-43.

①

②

③

Wife cake

Water dough
cake flour 120g
lard 30g
water 60g

Lard dough
cake flour 100g
lard 50g

Filling
candied winter melon 65g
sugar 45g
friedglutinous rice flour 50g
lard 30g
water 35g
candied osmanthus

Filling

1 Soak the candied winter melon in hot water until soft. Drain to make it less sweet.

2 Puree the candied winter melon in a food processor. Add all remaining filling ingredients. Knead into elastic dough ① ~ ②.

3 Divide into 6 equal pieces ③.

Crust, assembly and baking

1 Knead all water dough ingredients together until smooth and resilient.

2 Knead all lard dough ingredients together until the mixture resemble mashed potato.

3 Divide the water dough and lard dough into 6 equal pieces separately. Roll them round. Wrap a ball of lard dough in a piece of water dough. Seal the seam and roll it out into a long thin piece with a rolling pin. Roll it up along the length. Press it flat again gently. Fold in thirds from the sides toward the centre.

4 Roll each dough out into a thin round disc

5 Wrap a piece of filling in a piece of dough. Seal the seam and press it flat into a round patty ④. Let sit for 15 minutes. Brush egg yolk on top. Bake in a preheated oven at 220°C for 15 minutes. Serve.

* For the crust, follow steps 1-34 in "Basic puff pastry" recipe on p.40-43.

Buns

Buns are yeasted flour dough cooked after fermentation. They can be stuffed with filling or plain. They can be steamed, baked or fried in oil. Buns with filling include pan-fried pork buns and longevity peach buns. Buns without filling include plain steamed buns and silver thread rolls.

Bread fermentation has a long history including methods like alcohol fermentation, sourdough starter, bicarbonate rising, old dough and poolish, which are still used today. Generally speaking, we have more control on the fermentation these days and buns are the representative of the profound culture of yeasted bread.

Buns are made from flour, with water, yeast or preferment added. Other ingredients such as meat or vegetables are also used togive countless varieties of buns.

To begin, make the dough first. Mix together ingredients such as flour, water, yeast, sugar,grease and salt. To make the best buns, you should be picky about your flour. There are many kinds of flour out there and making the right choice is a crucial step for quality buns.

Also pay attention to the ratio of water to flour. You may also use other liquid instead of water. Just deduct the volume of liquid used from the volume specified for water.

Generally speaking, Cantonese buns are fluffier and sweeter than their Northern counterparts. Old dough is typically used and we prefer cake flour for its silky texture and snow-white colour. On the other hand, Northern Chinese prefer buns that are chewier in texture. That's why they use all-purpose flour and add preferment togive their buns more flavour.

Healthy steamed buns

Coconut milk buns

cake flour 300g

(or 270g cake flour+30g bread flour if you prefer more springy texture)

sugar ..... 35 to 40g

baking powder 4g

fresh yeast 10g

coconut milk (or water or milk) 40g

milk 130g

Yellow sweet potato buns

cake flour 220g

bread flour 30g

Taiwanese yellow sweet potato powder 50g

sugar ..... 35 to 40g

baking powder 4g

milk 175g

fresh yeast 10g

Purple sweet potato buns (Same as yellow sweet potato buns)

Replace Taiwanese yellow sweet potato powder with 50g Taiwanese purple sweet potato powder

Spinach buns

cake flour 300g

frozen spinach 100g
(thawed and puree with milk)

sugar ..... 35g to 40g

baking powder 4g

fresh yeast 10g

milk 80g

Black sesame buns (Same as coconut milk buns)

Replace coconut milk with 30g of Taiwan-eseground black sesame or black sesame paste

Milk 170g

Beetroot buns (Same as spinach buns)

Replace frozen spinach with 70g raw beetroot puree

Milk 100g

Carrot buns (Same as spinach buns)

Replace frozen spinach with 80g raw carrot puree

Milk 90g

Method

1 Measure all dry ingredients. Add half of the wet ingredients (including vegetable puree). Knead into marbling pattern. Then slow pour in the rest of the wet ingredients while kneading into a firm dough ① . Let it rest for 5 to 10 minutes. If you don't have a pasta roller, knead until smooth. Then roll it out with a rolling pin while trying your best to burst all air pockets. Fold the dough in thirds or quarters. Repeat the rolling step. If you have a pasta roller, you don't need to knead the dough as much ② . Even a rough dough works well after rolled through the pasta roller 3 or 4 times. The dough will turn out smooth and neat in shape.

2 Spray some water on the dough and roll it up into a log (the water helps adhesion). Then roll it into a long log with your desired thickness ③ . Bear in mind that the shape of the end product depends on the cross section of this dough log. When you slice the dough, use a sharp blade for neat cuts. Cut them into equal pieces. Do not slice them too thinly as they're likely to tip over.

3 A bamboo steamer works best for steaming buns because it picks up any excessive steam without forming condensation. Put the buns into the steamer and let them proof for 30 to 40 minutes. Make sure you leave enough space between the buns for expansion so that they don't stick together. On a warm day, just cover the lid and let them proof at room temperature. On a cold day, use warm water at 30°C to 40°C to make the dough. Cover the lid and put the steamer over a pot of hot water so that the buns proof at warmer temperature.

4 Steam the buns over medium heat for 14 minutes. Buns steamed over low heat may not be cooked through properly. They tend to be sticky and crumbly in texture without any elasticity. Buns steamed over strong heat tend to wrinkly on the surface and might even burst. Throughout the steaming process, open the lid once or twice to let the steam out. Or, you may cover the lid loosely so that excessive steam may escape. If you're using a metal steamer, put a clean towel over the buns to pick up the condensation. Otherwise, there may be blisters on the buns. Do not open the steamer right after the steaming time is up. Turn off the heat and leave them in the steamer for 2 more minutes before opening the lid. This step lets the buns cool down slowly without thermal shock. Otherwise, the buns may shrink and become too stiff.

Once youget agrasp on the basic plain buns, you may churn out a series of variations by adding all-natural ingredients and colours. These gorgeous and colourful buns aregreat for all ages and truly a healthy treat.

Tips

1 Do not let the dough rest for too long. Otherwise, the holes in the crumb will be too large.

2 After rolling the dough to your desired thickness, make sure you grab tightly when you roll it into a log. You must do it quickly so that the dough rise evenly.

3 After shaping, pay attention to the proofing time. Over or under-proofing may take its toll on the quality of the buns.

4 When you steam the buns, fine-tune the steaming time according to the heat of your stove.

5 You can use all-purpose flour to replace cake and bread flour.

Silver thread rolls with fried spring onion and pine nuts

Serves: makes 16 mini rolls

Dough
cake flour 500g
sugar 45g
baking powder 7g
fresh yeast 18g
water 140g
milk 140g

Filling
pine nuts 100g
sea salt 5g
chopped spring onion 4 to 5 tbsp
fried in goose fat

Method

1 Toast or bake the pine nuts till lightly browned. Grind them and mix with chopped spring onion fried in goose fat and sea salt.

2 Make the dough according to step 1 of the recipe "Healthy steamed buns" on p.190. Divide into 2 equal halves. Set one half aside.

3 Divide one piece of dough from step 2 into 2 equal pieces. Roll each out with a rolling pin. Spread chopped spring onion fried in goose fat mixture evenly on each piece.

4 Stack one layer of dough over the other ① . Spread chopped spring onion fried in goose fat mixture. Make a cut at the centre. Stack two layers over the other. Cut into thin strips evenly ② (see figure).

5 Smear the remaining chopped spring onion fried in goose fat mixture onto the thin strips of dough.

6 Take the remaining plain dough from step 2 and cut into quarters. Roll each piece out until long enough to wrap around 1/4 of the thin strips from step 5. Wrap thin strips into the plain dough ③ . Brush water on the seam and pinch to seal well. Put it on a counter with the seam down. Slice into equal pieces ④ . Put each on a piece of baking paper. Garnish with pine nuts. Steam over medium heat for 14 minutes.

Tips

In step 6, you may steam the rolls in whole without slicing them. Just slice before serving.

Longevity peach buns

Dough A

cake flour 600g

water ..... 280 to 300g

fresh yeast 10g

sugar 2g

Dough B

sugar 40g

lard 10g

baking powder 5g

Filling

white lotus seed paste or red bean paste 450g

water-soluble magenta food colourin

Method

1 Knead the dough A ingredients until smooth. Wrap it in cling film. Leave at room temperature (25°C to 32°C) for 45 to 60 minutes to rise until it doubles in volume. Conduct finger test for the dough. Dip your index finger into some flour. Poke the dough slowly with your finger. If the dough has risen enough, the hole will not change in size ① .

2 Add dough B ingredients. Keep kneading until the dough is smooth and doesn't stick to your hands. Wrap it in cling film and rest for 20 minutes.

3 Divide the dough into equal pieces weighing 25g each. Wrap some filling in each piece of dough. Seal the seam and roll one end pointy, the other end round. Line a bamboo steamer with baking paper ② . Arrange the buns over the paper. Leave the buns proof for the last time for 20 minutes. Steam for 12 minutes. Press the buns with your fingers to test their doneness. When the buns are cooked through, the mark of your finger should disappear as the buns bounces back.

4 When the buns are still hot, make an indentation in the middle with a bamboo skewer or the back of a knife so that they look like peaches ③ .

5 Dissolve the food colouring in some hot water. Dip a new toothbrush into the colour water. Splatter onto the buns from about 6 to 8 inches away by scraping the bristles with a paring knife towards yourself ④ .

Tips

If the pointy ends of the buns are not pointy enough after steamed, you can pinch them with your fingers.

Buns with date and osmanthus filling

Dough

cake flour 270g

bread flour 30g

sugar 30g

baking powder 5g

fresh yeast 10g

water 120g

milk 40g

Filling

snow dates from Tien Shan or large red dates 40

candied osmanthus ..... 3 tsp

rock sugar ..... 50 to 60g

water

Method

1 To make the filling, put the dates, rock sugar and candied osmanthus into a pot. Add enough water to cover (1cm above the ingredients). Boil the mixture until the dates are soft, but not mushy. Let the dates cool in the syrup until completely cool. Drain (and you may set aside the syrup for other uses) ① .

2 To make the dough, please refer to step 1 of the recipe "Healthy steamed buns" on p190.

3 Roll up the dough into a long log. Cut into pieces weighing 25g each. Press each flat. Pass it through a pasta roller or roll it into an oval with a rolling pin. Trim the edges. Put in a date. Fold the dough upward to wrap around the date, while exposing the top third of it. Pinch to shape the bottom neatly ② ~ ④ .

4 Put the buns into a bamboo steamer. Let them proof for 15 minutes. Make sure you leave enough space between the buns so that they won't stick together after expanding.

5 Steam the buns over medium heat for 6 minutes. Open the lid once to let the steam out halfway through the steaming process.

Pan-fried pork buns with fermented shrimp paste

Serves: makes 40 buns

Filling
Japanese cabbage 400g
sea salt
ground pork 600g
fermented shrimp paste 30g
sugar 20g
caltrop starch 8g

Plain dough
plain flour 400g
fresh yeast 5g
baking powder 2.5g
water 185g
lard 5g

Scalded dough
plain flour 200g
sea salt 5g
lard 25g
hot water (85°C to 100°C) 80g
cold water (to adjust consistency) 30g

Filling

1 Finely shred cabbage. Add sea salt and mix well. Leave it till the cabbage wilts. Rinse in water to remove the salt. Squeeze dry.

2 Season with pork with fermented shrimp paste and sugar. Mix well. Put into a mixing bowl and stir to mix well.

3 Add cabbage and caltrop starch. Mix well ① .

Plain dough

1 Mix the dry ingredients together. Add water (Don't pour all in at once. Save some to adjust the consistency.)

2 Knead into dough. Add oil and knead again.

3 Shape the dough into a boule. Put it on the counter with the seam side down. Let sit for 15 minutes. Meanwhile,make the scalded dough.

Scalded dough

Mix the flour, salt and oil together. Add boiling water and stir with chopsticks quickly. Wait till the mixture no longer sends off steam. Add cold water little by little to adjust the consistency. Knead until smooth.

Assembly

1 Knead the plain dough with the scalded dough till fully incorporated. Shape into a boule. Wrap in cling film. Let it sit for 20 minutes.

2 Roll the dough into a long log. Divide into 20-g pieces ② .

3 Press each piece into a flat round patty. Roll the edge thin with a rolling pin while leaving the centre thicker. Wrap in some filling and seal the seam.

4 Make a flour slurry by mixing 20g of flour with 300g of cold water.

5 Add oil into a non-stick pan. Heat the pan over low heat.

6 Arrange the buns side-by-side in the pan. Turn to medium-low heat to sear them ③ .

7 Pour in the flour slurry from step 4. The slurry should come up to 2/5 or half of the height of the buns. Cover the lid.

8 Cook until the slurry runs dry but the buns are still fluffy and elastic. Sprinkle with chopped spring onion and white sesames.

9 Fry until a crispy golden crust forms under the buns. Drizzle with some sesame oil.

10 Remove the pan from heat and let cool for 1 or 2 minutes. Carefully slide the buns out onto a serving plate without breaking them. Serve with spicy bean sauce and black Zhenjiang vinegar on the side.

①

②

③

Cakes and snacks

As opposed to Western cakes, Chinese cakes aren't necessarily fluffy, leavened or sweet. Every region in China has its own characteristic confections, cakes and sweet soups. On top of that, savoury snacks also make an indispensable part of Chinese diet. Chinese cakes are made with starch (such as rice flour, wheat flour or other starches) with liquid (such as water, eggs and oil) added. The batter is then steamed or baked till cooked. Chinese snacks, or better known as Dim Sum, are intricately crafted bite-size food that is filling somehow. These delicious munchies are served between proper meals to stop hunger pangs. They are more formal than snacks in the Western sense, but they aren't substantial enough to make a meal on their own. The common feature between Chinese cakes and snacks is their nostalgic and heart-warming nature. All ingredients are mundane and can be easily found in street-side stalls, such as sweet potato, taro, radish, starches and beans. They aren't exactly exotic or inventive. But they offer emotional comfort with their familiar taste and classic texture. It's back-to-basics food. It's real food with no pretence. It's a token of remembrance for our collective childhood memories.

Cakes and snacks come in countless types, and are made in countless ways. This book is not intended to be exhaustive, but rather serves as a sampler of what Chinese food has to offer. Owing to its central location, Hong Kong imports different ingredients from all around the world. That explains our unique Western twist on classic Chinese recipes, such as Tangyuan with cream cheese filling and mango mochi. Of course, traditional fare that we all grew up eating are also included, such as Put Chai Ko, radish cake and taro cake. These recipes are not difficult to learn but they aren't exactly easy to master. Just practice repeatedly to contemplate the secret tricks therein.

Snowy mooncake

Filling

dried split mung beans 225g

dried osmanthus 4g

unsalted butter 180g

milk 480g

sugar 150g

salted egg yolks (steamed and diced) 6

Method

1 Soak split mung beans and osmanthus in water separately for at least 4 hours. (Refrigerate it in summer so that they won't go stale).

2 Drain mung beans and osmanthus. Steam mung beans until soft.

3 Let cool the mung beans and blend in a blender.

4 Put unsalted butter and mung bean puree into a non-stick pan. Stiruntil hot. Slowly pour in milk and sugar while stirring.

5 Cook the mung bean mixture over low heat until thick. Pass it through a fine mesh to remove any lump ① .

Let cool. Stir in salted egg yolks. Divide into 40-g pieces. Roll each round and refrigerate

Crust

cake flour 25g

long-grain rice flour 50g

glutinous rice flour 60g

sugar 50g

milk 250g

condensed milk 35g

lard 15g

friedglutinous rice flour (for dusting)

Method

1 Melt lard in a double-boiler or over a pot of simmering water.

2 Mix all dry ingredients. Slowly pour in milk while stirring. Knead intosmooth dough.

3 Add condensed milk and lard from step 1.

4 Mix all ingredients. Pour into a cake tin. Steam over high heat for 35minutes.

5 Let cool. Do not drain the water and oil on top.

6 Put the steamed dough into a big mixing bowl. Knead until the oil and water incorporate well in the dough and the dough turns elastic ② .

7 Divide the dough into 20-g pieces. Wrap 40g of chilled mung beanfilling with each piece of dough.

8 Dust the mooncake mould with fried glutinous rice flour. Press the snowy mooncake into the mould and press firmly with your palm to leave a clear impression on it ③ . Gently tap the mould on the counter to turn the mooncake out. Or, you may use a springform mooncake mould. Refrigerate before serving.

Tips

1 This mooncake is not baked at last. Both the crust and the filling are cooked first before assembled and wrapped together. For food hygiene reason, do not handle the ingredients with bare hands after they've been cooked. Wear disposable gloves when you knead the crust, roll the filling or wrap in the filling.

2 Put a 1/3 or 1/4 of the mung beans into the blender at a time. Putting them in all at once may overload your blender. When you cook the mung bean filling, it's better to do it in a non-stick pan because it tends to burn easily. The filling should be creamy like peanut butter when hot. It will firm up after chilled.

3 You can make the filling and the crust dough ahead of time. They last in the fridge for a few days. After you wrap the filling in the crust, the mooncakes last in the freezer for up to 2 weeks.

4 To fill a regular two-tael wooden mooncake mould, the cake should weigh two Chinese taels, equivalent to 75g. The ratio of the crust to filling by weight should be 1 to 2, so that the crust should weigh 25g and the filling 50g. For this recipe, I used a 60-g springform mooncake mould. Each mooncake is made with 20g of crust and 40g of filling.

Date cake

Serves: makes 1 cake,
about 20cm X 20cm X 6cm

Ingredients
snow dates from Tien Shan 300g

red dates 300g

water 1kg

raw cane sugar slabs 150g

tapioca starch 450g

wheat starch 70g

Method
1 Put both dates into a bowl. Add water to cover. Steam for 3 hours. (Or cook them in an electric slow cooker). Pass the mixture through wire mesh. Then press the dates to squeeze out some liquid①. You'll need 1.15 kg of date soup for this recipe.

2 Put the date soup into a pot. Add raw cane sugar slabs and cook until they dissolve. If you find the soup too watery, cook it for a bit longer to evaporate the excessive water. The date flavour will also be stronger. Let cool.

3 Sieve tapioca starch and wheat starch into a bowl. Slowly pour in cooled date soup while stirring continuously until lump-free ② . Pass the mixture through a fine sieve.

4 Weigh the batter. Use 1/8 of the batter for each layer. Steam it before pouring another layer. As the starch tends to sink, you cannot pour in all batter at one time.

5 Grease a cake pan. Stir the batter well so that starch suspends in it. Pour in the first layer of batter ③ . Steam for 4 to 5 minutes. Pour in the second layer of batter. Repeat steaming and pouring in batter until all batter is used up ④ .

6 Let cool. Then refrigerate for a few hours until the cake is firm. Slice neatly into desired shapes. Steam again before serving.

Tips

1 After you steam the dates, do not press the dates too much. Otherwise, the date pour in all batter at one time. soup will be murky instead of clear.

2 Taste the date soup and adjust its sweetness by adding the right amount of sugar.

3 For the best result, use red dates from Tien Shan alongside regular red dates.

Coconut milk pudding with yellow split peas

Serves: makes 1 coconut pudding, about 20cm X 20cm X 6cm

Ingredients

dried yellow split peas 100g

store-bought fresh coconut milk ... 600g

(or use 1.9 kg homemade coconut milk and skip the water)

water ..... 1.3 kg

cornstarch 200g

sugar 280g

Method

1 Cook split peas in water until tender but without breaking down.

2 Add half of the water to cornstarch. Mix well into a slurry.

3 In a pot, heat the remaining water, coconut milk and sugar until steam appears without coming to a full boil.

4 Slowly pour in the cornstarch slurry from step 2 while stirring in one direction until it thickens and boils ① . Make sure you keep stirring to the bottom of the pot to prevent burning.

5 Put in the split peas ② . Bring to the boil and cook till it thickens. Turn off the heat.

6 Pour into cake tin ③ . Let cool and refrigerate.

Homemade coconut milk

1 Shell a coconut. Scrape off all the brown skin. You'd need 600g of coconut flesh and 600g of coconut water.

2 Put coconut flesh into a blender. For every 1 cup of coconut flesh, add 1.5 cup of water. Blend until fine. Strain the mixture and save the coconut milk. Add coconut water and mix well.

Tips

1 Scraping off the brown skin on the coconut flesh helps make the pudding whiter in colour.

2 You canget fresh coconut milk from spice stores or Indonesian grocery stores. It goes stale very easily and you must put it in freezer as soon as you can. Fresh coconut milk is high in fat content and you need to skim off some fat before using it to make dessert and snacks. Otherwise, the end product may be too greasy.

3 If you can't get fresh coconut milk from stores, you can make your own. Blend coconut flesh with water. Strain and add coconut water. The coconut water will give the pudding a richer and sweeter taste.

4 The amount of cornstarch is the key to this recipe. Adding too much will make the pudding too stiff. Make sure you stir patiently until the batter is cooked through. Otherwise, the pudding may taste floury.

5 If you use fresh coconut milk for this recipe, you don't need any more milk. It's because the colour of coconut milk will make the pudding white in colour.

 ①  ②  ③

Five-colour pumpkin and sweet potato cake

Ingredients

cornstarch 50g

long-grain rice flour 300g

cooked pumpkin 600g

water (depending on moisture content of pumpkin) 200-300g

coconut milk 250g

sugar 150g

raw purple sweet potato 200g

raw yellow sweet potato 200g

raw orange sweet potato 200g

raw taro 200g

Method:

1 Cut the pumpkin with skin into chunks. Remove the seed. Steam the pumpkin until soft. Mash the flesh while hot.

2 Dice all sweet potatoes and taro.

3 Put mashed pumpkin, sugar and 1/3 of the water into a pot. Bring tothe boil.

4 Mix rice flour, cornstarch and coconut milk with the remaining water.

5 Pour the boiling pumpkin syrup from step 3 into the flour slurry from step 4 ① . Stir immediately to mix well.

6 Add diced sweet potatoes and taro. Mix well ② ~ ③ .

7 Pour the mixture into agreased cake pan ④ . Steam for 90 minutes.

8 Let cool and slice. Steam until hot before serving.

Tips

The amount of water used depends on the moisture content of the pumpkin. Chinese pumpkin tends to be less starchy and more watery. Make the slurry with 200g of water if you use Chinese pumpkin. Japanese pumpkin is starchier and less watery. You may make the slurry with 300g of water instead.

Mango mochi

Serves: makes 2 mango mochi, about 20cm x 5cm x 4cm

Filling

Filipino mango puree 150g

coconut milk 35g

whipping cream 35g

sugar 50g

boiling water 75g

ice water 75g

gelatine leaf 9g

diced Filipino mango flesh

Crust

glutinous rice flour 200g

cornstarch 40g

sugar 100g

milk powder 40g

canned coconut milk 250g

milk 160g

water 200g

oil 40g

Topping

dried coconut shreds

Filling

1 Peel and stone the mango. Slice and puree it in a food processor or blender. Strain to remove any lumps.

2 Soak the gelatine leaf in cold water until soft.

3 Boil water and add sugar. Cook until it dissolves. Put in the gelatine leaf. Stir gently until it dissolves.

4 Add coconut milk, whipping cream and ice water. Add mango puree. Stir to mix well ① .

5 Pour the resulting mixture into moulds that have been rinsed with water. Put in diced mango flesh. Refrigerate until set ② .

Crust

1 Put all dry ingredients into a mixing bowl. Add sugar, milk, coconutmilk and water. Stir well. Add oil and stir again. Strain the mixture to remove any lump.

2 Pour the batter into a shallow square pan. Steam over high heat for 20 to 25 minutes until done. (Steaming time depends on the depth of the pan. It's done when it puffs up.) Let cool and set aside ③ .

Assembly

Cut the set mango filling to your desired sizes. When the crust dough has cooled completely, put on your disposablegloves. Grab a piece of dough. Dust your hands with some dried coconut shreds so that the dough won't stick. Press the dough into a long piece. Wrap in a piece of mango filling. Roll the mochi in dried coconut shred ④ . Shape the seam nicely. Refrigerate until chilled. Serve.

Tips

1 Just like making snowy mooncakes, the crust is cooked before you wrap in any filling. Thus, for food safety's sake, make sure you put on disposablegloves before you wrap mango filling in the mochi.

2 The crust dough should be cooled completely before you wrap in filling. Otherwise, the heat will melt the filling and it becomes difficult to handle.

3 Refrigerate the mochi till cold before slicing.

Put Chai Ko

(Hong Kong style steamed sweet rice cake in porcelain bowls)

Serves: makes 7 - 8 rice cakes

Ingredients
cooked red beans
1 raw cane sugar slab
long-grain rice flour 100g
wheat starch 22g
water chestnut starch 10g
water 500g
raw cane sugar slabs
(chopped finely) 120g

Method
1 Rinse raw red beans and soak in water for a few hours. Drain. Cook red beans in boiling water and add 1 slab of raw cane sugar. Cook until red beans are tender, yet without breaking down. Do not stir too much to keep the beans whole. Drain. Rinse off the starch.

2 In a mixing bowl, put 140ml of water, rice flour, wheat starch and water chestnut starch. Mix into slurry.

3 Put the remaining 360ml of water into a pot. Add chopped raw cane sugar. Cook until sugar dissolves.

4 Pour the boiling syrup from step 3 into the slurry from step 2 while stirring continuously ① . Put in some of the red beans ② .

5 Pour the resulting batter intogreased porcelain bowls ③ . Top with some red beans ④ .

6 Steam until done. Let cool slightly. Pierce the rice cake with two bamboo skewers. Turn it out of the bowl. Serve.

Tips

Wheat starch from different brands may yield rice cake of different firmness.

Healthful multigrain cake

> *Serves: makes 1 cake,*
> *about 23cm X 23cm X 6cm*

Ingredients

water chestnut starch 220g

tapioca starch 85g

water ..... 330g + 725g

sugar 155g

Multigrain mixture

dried red beans 120g

dried chick peas 220g

dried millet 60g

grits 100g

extra sugar for boiling beans

extra water for boiling beans

Method

1 Soak red beans and chick peas in water for 4 to 8 hours. (Keep in the fridge in summer to prevent them from going stale.)

2 Boil the red beans in water for 30 minutes. Drain to remove the bitter taste. Add water and some sugar. Cook until the red beans are tender. Drain and set aside.

3 Rinse the soaked chick peas. Cook them in some water and some sugar until tender. Drain and set aside.

4 Rinse millet and grits. Cook in water until tender. Drain and rinse under a running tap to make them less sticky. Squeeze out any water. Set aside.

5 Mix water chestnut starch and tapioca starch with 330g of water. Strain to remove any lump.

6 Put 155g of sugar into a pot and add 725g of water. Bring to the boil. Pour the water chestnut starch slurry from step 5 into the boiling syrup while stirring quickly. Cook until the mixture thickens. It's of the right consistency if it leaves no trace when you scoop it up with a ladle and pour it back into the pot ① .

7 Grease a cake pan.

8 Put the multigrain separately into bowls. Pour the batter from step 6 to just cover them ② .

9 Build layers of different grains in the cake pan. Pour in a layer of millet batter first. Swirl to coat the bottom of the pan ③ . Steam over high heat for 7 minutes.

10 Pour in a layer of red bean batter ④ . Steam over high heat for 10 minutes.

11 Pour in a layer of chick pea batter. Steam over high heat for 10 minutes.

12 Top with a layer of grits batter. Steam over high heat for 7 minutes.

13 Let cool and refrigerate until firm. Slice and serve chilled.

Tips

1 Make sure you keep stirring while pouring the slurry into the syrup. Otherwise, the starch may sink and the cake may turn lumpy.

2 Slice the cake only after you refrigerate it long enough. Otherwise, it's hard to slice it neatly.

3 You may use othergrains of your choice. Just adjust the cooking time according to the grain sizes.

4 The cake will look better if you choose grains in different colours.

5 You may cook a whole batch ofgrains and keep the leftover in the freezer after rinsing. They last up to a few months. Just thaw and use them in your next batch of cake.

 ①
 ②
 ③
 ④

Tangyuan with cream cheese filling coated in ground peanuts and coconut shreds

Filling

cream cheese 80g
 (or any soft cheese with neutral flavour)

sugar 40g

toasted peanuts (ground) 40g

toasted sesames

Crust

glutinous rice flour 100g

wheat starch 10g

sugar 20g

warm water 80g

lard 12g

Topping

dried coconut shreds 150g

sugar 40g

toasted peanuts (ground) 80g

Method

1 Mix all filling ingredients together. Refrigerate till firm. Divide into 8 equal pieces, each weighing about 18g ① .

2 Knead all crust ingredients together into smooth dough. Divide into 8 equal pieces, each weighing about 25g.

3 Mix all topping ingredients together. You may use more or less any ingredients according to your own taste.

4 Wrap 1 piece of filling in 1 piece of crust dough ② . Seal the seam and roll it round. Boil the Tangyuan in water for about 8 minutes or until they float. Remove from the water with a strainer ladle. Roll them in the topping to cover evenly③. Transfer onto a serving plate. Serve.

Coconut New Year cake

Serves: makes 1 New Year cake,
about 23cm X 23cm X 6cm

Ingredients

Indonesian palm sugar 400g

water 250g

glutinous rice flour 530g

wheat starch 80g

coconut milk 450g

Method

1 Cook the palm sugar in water until it dissolves. Set aside to let cool.

2 In a mixing bowl, put inglutinous rice flour and wheat starch. Add coconut milk slowly and stir into a paste. Then pour in the syrup from step 1 slowly ① while stirring continuously. Make sure there's no lump in the batter. Pass the batter through a fine mesh to remove any lump ② .

3 Pour the batter into agreased cake pan ③ ~ ④ . Steam until the cake turns slightly darker in colour. Check its doneness by inserting a bamboo skewer at the centre. It's done if it comes out clean.

Tips

Wheat starch makes the cake less sticky and firmer in texture. If you prefer more stickiness, you may omit wheat starch.

Radish cake with dace, fish stock and dried shrimps

> *Serves: makes 1 radish cake,*
> *about 26cm X 26cm X 6cm*

Ingredients

large dried shrimps 80g

white radish 2kg

minced dace (seasoned) 500g

coriander (diced)

spring onion (diced)

oil

2 cubes concentrated fish stockgel

long-grain rice flour 450g

cornstarch 70g

ground white pepper

sesame oil

sea salt 25g

Water 1.2kg

Method

1 Dice the dried shrimps. Peel the radish and grate into fine strips.

2 Stir together minced dace, coriander and spring onion.

3 Fry the minced dace into a patty in a little oil ① . Let cool and dice it.

4 Cook white radish and dried shrimps in 2/3 of the water until soft.Add stockgel and diced dace patty. Bring to the boil ② .

5 Add rice flour, cornstarch,ground white pepper, sesame oil and seasalt to the remaining water. Mix well.

6 Pour the rice flour slurry from step 5 into the boiling radish soupfrom step 4 ③ . Stir immediately to mix well.

7 Pour the batter into agreased cake pan ④ . Steam for 90 minutes.

8 Let cool and cut into large cubes. Fry in oil untilgolden. Serve.

Tips

Rice flour of different brands maygive the cake different consistencies after steamed. You may have to experiment a little on the proportion between rice flour and cornstarch for the consistency you prefer.

Savoury New Year cake in Hakka style

Ingredients

medium-sized dried
shiitake mushroom 6

dried shrimps 80g

pork shoulder butt
(with some fat) 300g

shallot 180g

oil

premium light soy sauce ..... 2 tsp

sugar 75g

glutinous rice flour 600g

long-grain rice flour 300g

ground white pepper

sesame oil

sea salt 25g

chicken bouillon powder 8g

water 800g

Method

1 Soak shiitake mushrooms in water until soft. Drain and dice them. Soak dried shrimps in water until soft. Rinse and set aside. Slice pork thickly. Rub salt, sugar and cornstarch on it. Leave for a while. Slice the shallot.

2 Stir fry shallot in oil untilgolden. Put in dried shrimps and shiitake-mushrooms. Stir until fragrant. Put in the pork and soy sauce. Stir until lightly browned ① .

3 In a mixing bowl, put inglutinous rice flour, long-grain rice flour,ground white pepper, sesame oil, sugar, sea salt and chicken bouillon powder. Mix well. Add water and knead into dough ② .

4 Put in the fried pork mixture from step 2. Knead well ③ .

5 Put the dough into agreased cake pan. Steam for 90 minutes.

6 Let cool and slice. Fry in a little oil untilgolden. Serve.

Taro cake with preserved meat and sausage

> *Serves: makes about 3 kg of*
> *taro cake*

Ingredients
Chinese preserved pork sausage ... 80g
Chinese preserved pork belly 80g
dried shrimps 15g
taro 450g
water 1.75kg
long-grain rice flour 338g
cornstarch 20g
shallot 2
dried scallops 15g

Seasoning
salt 20g
sugar 38g
chicken bouillon powder 10g
five-spice powder 9g
ground white pepper

sesame oil

light soy sauce

Method

1 Blanch the preserved pork sausage and pork belly in boiling water. Dice and set aside. Soak dried shrimps and dried scallops in water until soft. Set aside ① .

2 Dice the taro. Fry in oil briefly. Drain. Cook in 2/3 of the water and bring to the boil.

3 In a large bowl, mix cornstarch, rice flour and seasoning with theremaining water. Mix well.

4 Heat a wok and add oil. Stir fry shallot until fragrant. Discard theshallot. Set aside some diced preserved pork sausage and pork belly as garnish. Fry the remaining preserved pork sausage and pork belly, dried scallops and dried shrimps until fragrant.

5 Cook taro until soft. Put in the fried sausage mixture from step 4. Bring to the boil ② .

6 Stir the rice flour slurry from step 3 well ③ . Pour slurry into the boiling taro mixture from step 5. Stir well ④ . Pour the resulting batter into agreased cake pan. Steam for 90 minutes until cooked through. Arrange the diced preserved pork sausage and pork belly you set aside in step 4 on top.

7 Let cool. Refrigerate until firm. Slice and fry in some oil until golden. Serve.

Tips

1 When you dice the taro, you may finely chop some of it. Such taro bits will melt in the cooking process and add to the flavour of the cake.

2 You may use more or less five-spice powder according to your own taste.

3 The five-spice powder in packets that you get from wet markets usually tastes better than those bottled ones from supermarkets.

4 Similar to making radish cake, the rice flour slurry is poured into the boiling liquid which instantly thickens into a batter. The taro will suspend in the batter this way instead of floating on top.

①

②

③

④

Homestyle sponge cake

Serves: makes 26cm cakes

Preferment

long-grain rice flour 100g

cake flour 100g

fresh yeast 10g

water 120g

sugar 20g

Main dough

long-grain rice flour preferment 600g

plain flour 100g

vegetable oil 200g

water 300-400g

Syrup

raw cane sugar slabs 600g

water 400g

Method

1 Mix all preferment ingredients together. Knead to mix well. Leave it in a warm place (26°C-28°C) for 7 to 10 hours.

2 Mix all main dough ingredients together. Knead well. Leave it in a warm place (25°C-28°C) for 4 hours.

3 To make the syrup, cook raw cane sugar slabs in water until it dissolves. Pass it through a fine mesh and set the syrup aside.

4 Slowly add syrup to the dough ① . Stir with your hand to make a thin batter ② . Strain to remove any lump. Leave it in a warm place (32°C-35°C) for 4 to 5 hours until it bubbles ③ . Stir well ④ . Pour the batter into agreased cake pan. Steam for 90 minutes. Slice and serve.

Tips

According to my mom, you can only steam one cake in one steamer or wok. Otherwise, the cake won't turn out smooth and pretty. Besides, you should also pay attention to the steaming time. Make sure you've steamed it long enough. As opposed to other cakes, if it's undercooked the first time around, it will never cook properly no matter how many times you steam it.

流行中式點心

茶粿、酥餅、糕點、包子饅頭　一次學會

作　　者	獨角仙
編　　輯	徐詩淵
美術設計	林采瑤（美果視覺設計）
校　　對	徐詩淵、鄭婷尹

發 行 人	程安琪
總 策 畫	程顯灝
總 編 輯	呂增娣
主　　編	翁瑞祐、羅德禎
編　　輯	鄭婷尹、黃馨慧
美術主編	劉錦堂
美　　編	曹文甄
行銷總監	呂增慧
資深行銷	謝儀方
行銷企劃	李昀

發 行 部	侯莉莉
財 務 部	許麗娟、陳美齡
印　　務	許丁財
出 版 者	橘子文化事業有限公司

總 代 理	三友圖書有限公司
地　　址	106 台北市安和路 2 段 213 號 4 樓
電　　話	(02) 2377-4155
傳　　真	(02) 2377-4355
E - m a i l	service@sanyau.com.tw
郵政劃撥	05844889 三友圖書有限公司

總 經 銷	大和書報圖書股份有限公司
地　　址	新北市新莊區五工五路 2 號
電　　話	(02) 8990-2588
傳　　真	(02) 2299-7900

製　　版	興旺彩色印刷製版有限公司
印　　刷	鴻海科技印刷股份有限公司

初　　版	2017 年 5 月
定　　價	新臺幣 450 元
I S B N	978-986-364-102-5（平裝）

SAN Yau
http://www.ju-zi.com.tw
三友圖書
友直 友諒 友多聞

國家圖書館出版品預行編目 (CIP) 資料

流行中式點心：茶粿、酥餅、糕點、包子饅頭一
次學會 / 獨角仙作. -- 初版. -- 臺北市：橘子文化，
2017.05
　面；公分
ISBN 978-986-364-102-5(平裝)
1. 點心食譜 2. 中國
427.16　　　　　　　　　　　　106006229

地址： ＿＿＿＿縣/市 ＿＿＿＿鄉/鎮/市/區 ＿＿＿＿路/街

＿＿段 ＿＿巷 ＿＿弄 ＿＿號 ＿＿樓

三友圖書有限公司 收

SANYAU PUBLISHING CO., LTD.

106　台北市安和路2段213號4樓

親愛的讀者：
感謝您購買《流行中式點心：茶粿、酥餅、糕點、包子饅頭一次學會》一書，為感謝您對本書的支持與愛護，只要填妥本回函，並寄回本社，即可成為三友圖書會員，將定期提供新書資訊及各種優惠給您。

姓名 ＿＿＿＿＿＿＿＿＿＿＿＿＿＿ 出生年月日 ＿＿＿＿＿＿＿＿＿＿＿＿＿＿
電話 ＿＿＿＿＿＿＿＿＿＿＿＿ E-mail ＿＿＿＿＿＿＿＿＿＿＿＿＿＿＿＿
通訊地址 ＿＿＿＿＿＿＿＿＿＿＿＿＿＿＿＿＿＿＿＿＿＿＿＿＿＿＿
臉書帳號 ＿＿＿＿＿＿＿＿＿＿＿＿＿＿＿＿＿＿＿＿＿＿＿＿＿＿＿
部落格名稱 ＿＿＿＿＿＿＿＿＿＿＿＿＿＿＿＿＿＿＿＿＿＿＿＿＿

1 年齡
□ 18 歲以下　□ 19 歲～ 25 歲　□ 26 歲～ 35 歲　□ 36 歲～ 45 歲　□ 46 歲～ 55 歲
□ 56 歲～ 65 歲　□ 66 歲～ 75 歲　□ 76 歲～ 85 歲　□ 86 歲以上

2 職業
□軍公教 □工 □商 □自由業 □服務業 □農林漁牧業 □家管 □學生
□其他 ＿＿＿＿＿＿＿＿＿＿＿＿＿＿＿＿＿＿＿＿

3 您從何處購得本書？
□博客來　□金石堂網書　□讀冊　□誠品網書　□其他＿＿＿＿＿＿＿＿＿
□實體書店 ＿＿＿＿＿＿＿＿＿＿＿＿＿＿＿＿＿＿

4 您從何處得知本書？
□博客來　□金石堂網書　□讀冊　□誠品網書　□其他＿＿＿＿＿
□實體書店 ＿＿＿＿＿＿＿ □FB（三友圖書 - 微胖男女編輯社）
□三友圖書電子報　□好好刊（雙月刊）　□朋友推薦　□廣播媒體＿＿＿＿＿＿＿

5 您購買本書的因素有哪些？（可複選）
□作者 □內容 □圖片 □版面編排 □其他 ＿＿＿＿＿＿＿＿＿＿＿

6 您覺得本書的封面設計如何？
□非常滿意 □滿意 □普通 □很差 □其他 ＿＿＿＿＿＿＿＿＿＿＿

7 非常感謝您購買此書，您還對哪些主題有興趣？（可複選）
□中西食譜 □點心烘焙 □飲品類 □旅遊 □養生保健 □瘦身美妝 □手作 □寵物
□商業理財 □心靈療癒 □小說 □其他 ＿＿＿＿＿＿＿＿＿＿＿

8 您每個月的購書預算為多少金額？
□ 1,000 元以下　□ 1,001 ～ 2,000 元□ 2,001 ～ 3,000 元□ 3,001 ～ 4,000 元
□ 4,001 ～ 5,000 元□ 5,001 元以上

9 若出版的書籍搭配贈品活動，您比較喜歡哪一類型的贈品？（可選 2 種）
□食品調味類　□鍋具類　□家電用品類　□書籍類 □生活用品類　□DIY 手作類
□交通票券類　□展演活動票券類 □其他 ＿＿＿＿＿＿＿＿＿＿＿

10 您認為本書尚需改進之處？以及對我們的意見？
＿＿＿＿＿＿＿＿＿＿＿＿＿＿＿＿＿＿＿＿＿＿＿＿＿

感謝您的填寫，
您寶貴的建議是我們進步的動力！